SALMON:
ECONOMICS AND MARKETING

SALMON: ECONOMICS AND MARKETING

SUSAN SHAW,
Senior Lecturer,
Department of Business and Management,
University of Stirling
and
JAMES F. MUIR,
Senior Lecturer,
Institute of Aquaculture,
University of Stirling,
Scotland

CROOM HELM
London & Sydney

TIMBER PRESS
Portland, Oregon

Softcover reprint of the hardcover 1st edition 1987
Croom Helm Ltd, Provident House, Burrell Row,
Beckenham, Kent, BR3 1AT

Croom Helm Australia, 44-50 Waterloo Road,
North Ryde, 2113, New South Wales

British Library Cataloguing in Publication Data

Shaw, Susan
    Salmon: economics and marketing.
    1. Salmon-fisheries 2. Salmon industry
    3. Fish trade
    I. Title II. Muir, James F.
    338.3'72755   SH346
ISBN-13: 978-94-010-7925-9

First published in the USA 1987 by
Timber press,
9999 S.W. Wilshire,
Portland, OR 97225,
USA

ISBN-13: 978-94-010-7925-9   e-ISBN-13: 978-94-009-3177-0
DOI: 10.1007/978-94-009-3177-0

CONTENTS

LIST OF TABLES AND FIGURES

Figures

Preface and Acknowledgements

The past decade has been an exciting one in the
development of the international salmon business
because of the increased richness of the available
salmon resources.   Increased supplies have made
it possible to develop new markets and to allow
more people to eat salmon.   These increases arise
in part from a high level of natural abundance of
Pacific salmon and in part from our increased
ability to control salmon production by farming
and enhancement activities.
    Fishing, farming and the enhancement of
natural stocks are being carried out in the
Atlantic, the Pacific and in both Northern and
Southern Hemispheres.   The resulting salmon
products meet and compete with each other in many
markets throughout the world.   Although the
degrees of competition vary between species and
between markets, all salmon business is affected
by such market and supply interrelationships.
The purpose of this book is to attempt to explain
some of the interactions between fishing and
farming, between different markets, between
different species and between the production costs
of different systems.
    There are many gaps in this book - in many
areas statistical data is inadequate and the
amount of published research which has been
carried out on processing and marketing, with the
exception of North America, is very small.   A
further constraint is that the picture is changing
rapidly through the combined effects of changing
markets, changing supplies and changing technologies.   The authors hope, nevertheless, that
this book will help the interested reader to
understand something of the complex influences and

processes which go together to make this a fascinating industry to study.

We have many debts. We have received considerable help from a great many people - from harvesters, dealers and processors to industry associations, government officials and academic researchers, all of whom gave their time most generously and without whom it would have been impossible to write this book. Particular debts are owed to government departments - the Department of Fisheries and Oceans in Canada, the National Marine Fisheries Service in the United States, the Department of Agriculture and Fisheries for Scotland and to the North Atlantic Salmon Conservation Organisation for their help.

We are grateful for the financial support of the Carnegie Trust for Scottish Universities which made it possible for one of the authors to visit Alaska in 1984. The Overseas Development Administration of the British Government gave generous permission for us to use unpublished material in a study of salmon ranching which we carried out for them (with G Poulter and R Morgan) in 1983. Our departments at Stirling University - the Department of Business and Management and the Institute of Aquaculture - gave us both valuable support. Jenny Rana acted as research fellow for part of the period and we are very grateful for her efficient assistance. We would like to thank Ann Cowie and Shirley Hewitt for their help in preparing the manuscript and Elise Macrae for her line drawings. Needless to say, any errors are the responsibilities of the authors alone.

Chapter One

INTRODUCTION

1.1 OUTLINE OF THE BOOK

The salmon family is among the world's most well
regarded and highly priced fish as well as being
among the most international. From harvesting
points far remote from each other different types
of salmon meet and compete in the major markets of
the world. Salmon are harvested along the Pacific coast of the Soviet Union, along the West
coast of North America and from the Eastern Coast
of North America to the fjords of Norway. Increasingly in recent years harvests are also coming from the waters of the Southern Hemisphere.
In turn salmon harvests are marketed widely in
fresh, frozen and smoked form throughout the world
through marketing systems of considerable complexity and variety. This book is an attempt to
describe this international business system, to
identify the ways in which it has been changing
and to indicate the directions of future change.
The analysis follows salmon from harvesting,
through processing, to the final consumer. In
order to set the scene for subsequent analysis and
discussion, the remainder of this first chapter
provides a brief description of the fish itself,
its life cycle, methods of fishing and rearing and
a description of the main processed salmon products. The differences between the species are
discussed in some detail, since, as will be seen
later, each has different production characteristics and a different position in the market
place.
    Chapter 2 describes the main patterns of
supply and the main trading patterns for salmon
and salmon products. Chapters 3, 4 and 5 look at
the supply side in more detail and are concerned

with the issues involved in the management of salmon fishing, farming and ranching and the implications for production costs and the availability of supplies. Chapters 6 and 7 investigate processing, marketing and distribution and Chapter 8 discusses the workings of salmon markets and the main factors affecting the demand for salmon. Finally, chapter 9 summarises the main findings of the study and considers the issues likely to define the future of the industry. The main interest of the book is in salmon sold commercially and mention is only made in passing of the rather different, though important, issues involved in the management of sport or recreation fisheries.

This book does not cover all species of salmon but concentrates on the main species of current commercial significance. These are the five commercially important species of Pacific salmon all of which belong to the single separate genus ONCORHYNCHUS together with SALMO SALAR, the Atlantic salmon. The related SALMO GAIRDNERI (rainbow trout) and SALMO TRUTTA (brown trout/seatrout) are outside the scope of the book, although since these compete closely with salmon, the relationship is discussed.

1.2 THE SALMON: LIFE CYCLE AND SPECIES

The family salmonidae is indigenous to the Northern Hemisphere and is found from the temperate zone northwards to beyond the Arctic Circle, in both Pacific and Atlantic waters. They are not native to the Southern Hemisphere but have been successfully introduced into New Zealand, into Argentina, Venezuela and more recently Chile and the Kerguelen Islands. Although the different members of the salmon family have different physical and life cycle characteristics, they have certain basic features in common. Spawning occurs in freshwater, sometimes in lakes, but usually in gravel reaches of streams or rivers with free running water. The usual spawning season is in autumn but a significant stock spawn in late spring. Fertilised eggs then develop in the gravel substrate for a period of several months, usually over the winter. After hatching as "alevin" the young fish feed initially on an attached yolk sac until they reach the "fry" stage when they start to

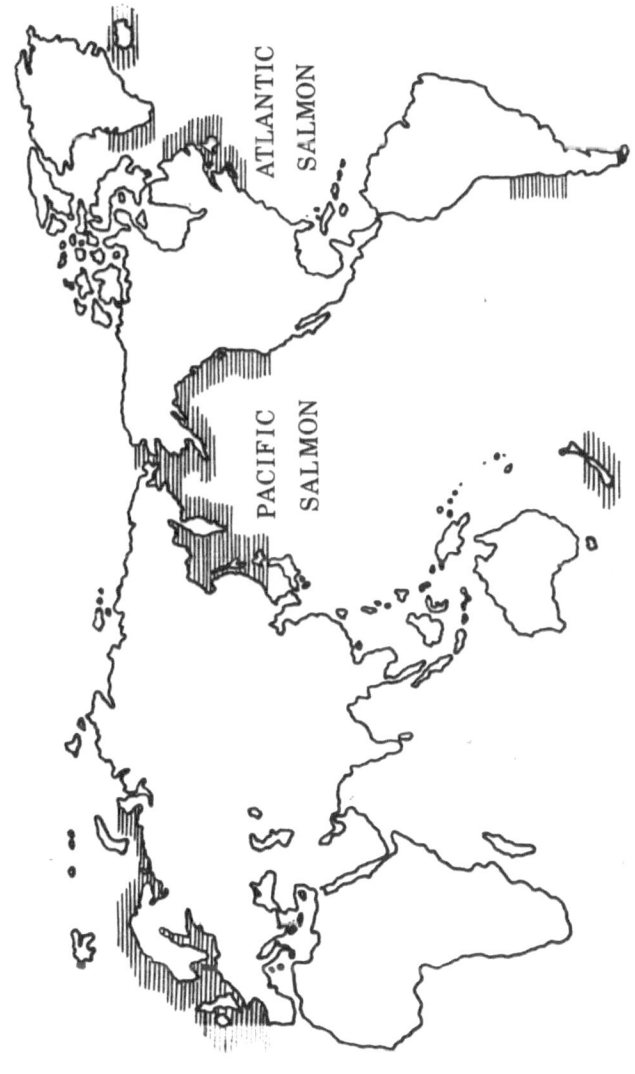

Figure 1.1:   Salmon harvesting points

ATLANTIC SALMON

PACIFIC SALMON

Figure 1.2: The salmon life cycle

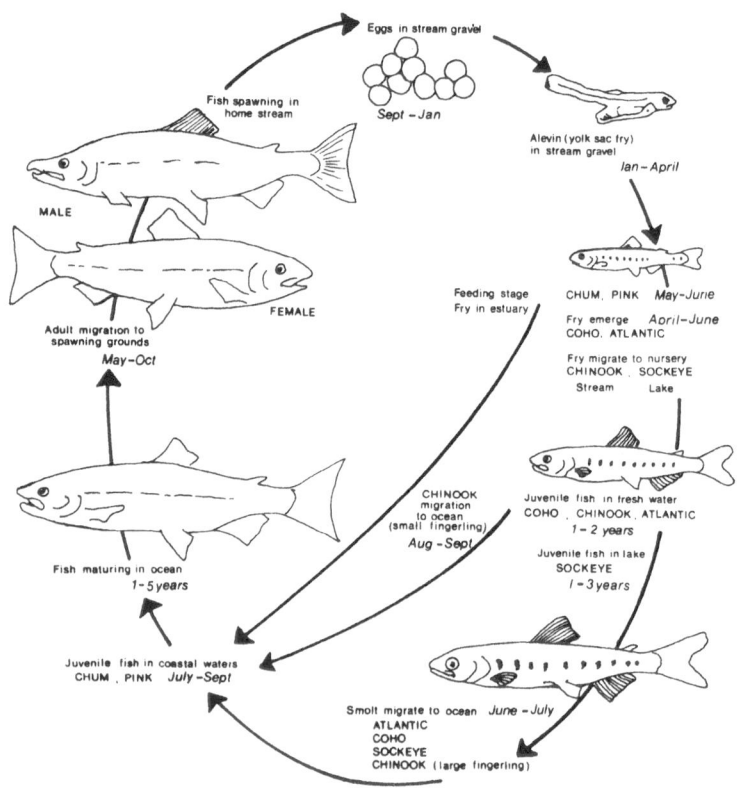

Eggs in stream gravel

Sept – Jan

Alevin (yolk sac fry)
in stream gravel
Jan – April

Fish spawning in
home stream

MALE

FEMALE

Adult migration to
spawning grounds
May – Oct

Feeding stage
Fry in estuary

CHUM, PINK   *May – June*

Fry emerge   *April – June*
COHO, ATLANTIC

Fry migrate to nursery
CHINOOK , SOCKEYE
    Stream   Lake

CHINOOK
migration
to ocean
(small fingerling)
Aug – Sept

Juvenile fish in fresh water
COHO , CHINOOK , ATLANTIC
1 – 2 years

Juvenile fish in lake
SOCKEYE
1 – 3 years

Fish maturing in ocean
1 – 5 years

Juvenile fish in coastal waters
CHUM , PINK   *July – Sept*

Smolt migrate to ocean   *June – July*
    ATLANTIC
    COHO
    SOCKEYE
    CHINOOK (large fingerling)

feed externally and move in the open water. Some
salmon go to sea as fry while others remain in
freshwater for longer periods until they reach the
"smolt" stage at which they become silvery in
appearance. Both fry and smolts are small: fry
weigh about half a gramme in early stages, about
5-10 grammes at seaward migration and smolts 30 to
50 grammes.
     Different species spend varying periods at sea
en route to and from feeding grounds but most
salmon are anadromous, i.e. at the end of the sea
feeding stage they return to their native rivers
to spawn. Some, as yet commercially insignifi-
cant, stocks remain in freshwater as "landlocked"
salmon. Pacific and most Atlantic salmon die
after spawning but some Atlantic salmon complete a
second and very occasionally a third cycle before
dying. The strength of the homing ability varies
but most find their way back to the rivers from
which they emerged. This migration, from sea
feeding grounds back to freshwater spawning
grounds can involve travel of enormous distances
since sometimes sea feeding grounds are thousands
of kilometres from estuary mouths. Even when
salmon return to freshwater the distances can be
considerable. For instance, some stocks of Yukon
river chinook travel 3000 kilometres from entry to
the estuary to their spawning grounds high up the
river. On this journey they climb to an altitude
of over 700 metres above sea level.

1.3 ATLANTIC SALMON (SALMO SALAR)

This species, less abundant than Pacific salmon,
occurs naturally in Atlantic waters but has been
introduced to a limited extent in the Pacific and
Southern Oceans. At the southern end of their
range they spawn in freshwater in the autumn and
winter and migrate to the sea in the spring, one
year after hatching. Salmon from colder more
northerly rivers may take 5 to 6 years to reach a
sufficient size for migration; the overall
average is 3 years. Salmon from the North
American and European coasts meet and mingle in
the feeding grounds of the North Atlantic. The
sea feeding areas are on the edge of the Arctic
ice pack, off Greenland and in the Barents and
Norwegian seas. Some Atlantic salmon termed
grilse mature early and return to spawn after only

one winter at sea when they weigh between 1.5 and
3.5 kilos. For the majority, returning after a
second winter at sea, the weight range is 3 to 6
kilogrammes; for fish returning after a third
winter the weight range is 5 to 14 kilogrammes.
Atlantic salmon are bright silver when at sea,
darkening on return to the rivers. The flesh is
dark pink with a high fat content and Atlantic
salmon have a premium position in the market
place, especially in Europe. They are not canned
but are sold mainly in fresh form as whole fish,
steaks and fillets and also as smoked salmon.

1.4 THE PACIFIC SALMONS

The five main species of Pacific salmon are native
to the rivers flowing into the North Pacific and
Eastern Arctic Oceans, the Bering Sea and the Sea
of Okhotsk. Some species have been introduced
into the Atlantic and Southern Oceans. They all
occur naturally in large quantities (details of
landings by country can be found in Chapter 2).
There are two other members of the ONCORHYNCHUS
family: ONCORHYNCHUS MASU which occurs in relatively small quantities in Japan and off the coast
of Manchuria and ONCORHYNCHUS RHODURUS which
occurs in even smaller quantities off southern
Japan.
Within each of the species described below
there are variations in the stock characteristics
and quality from different river systems. Knowledgeable buyers are aware of this and will seek
information on the precise source of salmon when
purchasing.

1.4.1   Chinook (Oncorhynchus tshawytscha)
(also known as Kings, Springs, Tyee and Quinnat)
These are the least abundant but the largest of
the Pacific salmon and are found mainly in Alaska,
off the Kamchatka Peninsula and in the Bering Sea,
but they have also been introduced into the
Southern Hemisphere. They spawn in the period
April to September. Some descend to sea as fry
in the following summer but others stay in freshwater for 1 to 2 years before going to sea as
smolts. They return after up to 6 or 7 years
(the largest fish) but most of those going out as
fry return after about 5 years of sea feeding.
On return they weigh 2 to 20 kilogrammes but can
reach much larger sizes: the largest fish was

Figure 1.3:   The salmons

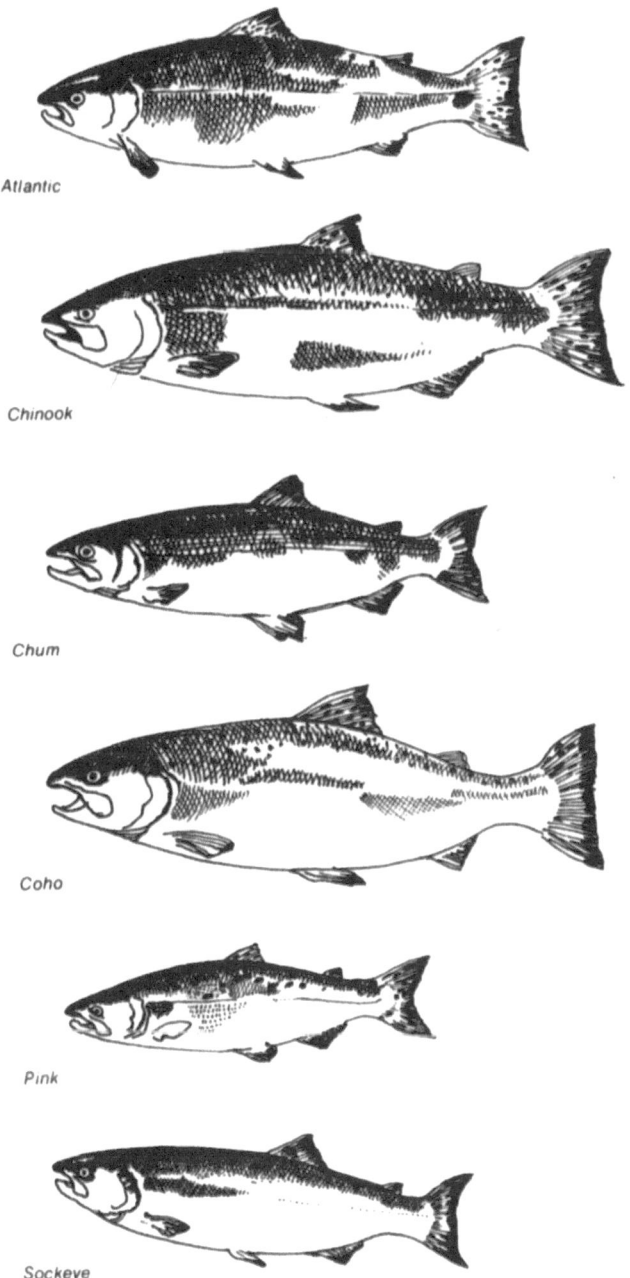

Atlantic

Chinook

Chum

Coho

Pink

Sockeye

recorded in 1949 in Alaska at 57 kilogrammes. Typical commercial sizes are 4 to 10 kilogrammes.

Colour and condition change as the fish move from open waters towards their native rivers, although this is somewhat less for chinook than some other species. When chinook are caught in open waters ("ocean brights") they have a silver skin and a deeper body than other species. Their backs are dark green/blue black with small black spots and the belly is silver white. As they move closer inshore ("semi brights") there is a slight dulling of colour and when they enter freshwater they have become black on the back and brown on the sides.

Chinook, particularly "brights", are regarded as premium quality fish. The flesh is red, firm and has a high fat content. Chinook are much sought after for smoking, freezing and eating fresh. Only small quantities of chinook are canned. A small number of chinook are known commonly as "white kings" because of their white flesh and while of equally good quality they obtain a rather lower price than "red kings". Chinook are primarily sold in slices, steaks and fillets.

1.4.2   Chum (Oncorhynchus keta)
(also known as Dog salmon)
Chum are found in large quantities in Asia, North America and Japan. They normally spawn in the period September to January although it may be somewhat earlier than this in the far north. They hatch after 120 days and migrate to the sea as fry in the following summer. On average they reach maturity and return to freshwater after 3 years although it may vary between 1 and 6 years. At 3 years the average weight is 2.5 kilogrammes and 6 kilogrammes is. the largest weight recorded. Commercial sizes range from 2 to 5 kilogrammes.

Ocean caught fish (ocean bright) are a dark metallic greenish blue on top becoming silver on the sides with fine pale bars sometimes present. As they move closer to the rivers (semi brights) dark bars appear on the sides and the skin on the back becomes duller. Dark chums have well developed dark bars on deep red sides. With all species the value of the salmon falls as it becomes duller but this is particularly noticeable in the case of chum.

Chum are medium priced salmon with a firm pink flesh and moderate fat content. They are sold fresh and frozen and are canned and smoked. Fresh and frozen chum are sold whole and in steaks and fillets.

1.4.3   Coho (Oncorhynchus kisutch)
(also known as silvers)
These are found in large numbers in North America, off the Kamchatka Peninsula and in the Bering Sea. Although not native to the region, they have in addition been successfully introduced into the Southern Hemisphere. This is the Pacific species whose behaviour and life cycle is closest to that of Atlantic salmon. Normally they spawn in the autumn or early winter, the fry emerging in April or May and remaining in freshwater for 1 to 2 years before leaving for the sea as smolts. They remain at sea for 2 years and at the end of their marine life they weigh between 2.5 and 6 kilogrammes. Smaller quantities of male "jacks", the equivalent of grilse, return after 1 year at typical weights of 1 to 3 kilogrammes.
Ocean caught "bright silvers" are dark metallic blue on the back and upper sides with a few small black spots on the silver body. "Semi brights" have a slight pink shading along the belly while "dark silvers" have red skin with darker backs and a pronounced hook to the mouth. The flesh is orange red with a firm texture and delicate flavour. They have a high fat content and good colour retention when cooked. Only small quantities are canned and coho is more frequently sold in either fresh, frozen or smoked form and in steaks, fillets or whole.

1.4.4   Pinks (Oncorhynchus gorbuscha)
(also known as Humpback or Humpy)
Pinks are widespread in large rivers in North America, Asia and in the USSR. Some small numbers can also be found in rivers along the North Atlantic coastlines of the USSR as a result of earlier stock transplantings. They spawn in late summer not far above the head of the tide and descend to sea as fry in the following spring. They spend 14 to 16 months at sea and reach sexual maturity in 2 years at which time their average weight is 2-3 kilogrammes, although some may weigh up to 4 kilogrammes. As the fish hold to a

distinct 2 year cycle, stocks on alternate years are distinct and have different stock level patterns.

Ocean bright pinks have a slim body and shiny silvery skin with small scales and large black spots on head and tail. Male semi brights have a slight hump. Dark pinks have a much darker body with coarser skin and the male has a pronounced hump, hence the popular name. The flesh is firm and rose coloured and is of a fine texture and delicate flavour. Pinks must be carefully and quickly handled when caught because the flesh is soft and easily damaged. They are canned in large quantities and sold fresh and frozen, often in whole form if they are smaller fish. Only small quantities are smoked. Net caught pinks typically sell at the lower end of the price range for salmon, but in the last few years increasing quantities of pinks have been caught by troll boats (see below). Since the pinks caught by this method are of higher quality they have sold for better prices.

1.4.5   Sockeye (Oncorhynchus nerka)
(also called Red and Blueback)
Sockeye are found in North America particularly Alaska and British Columbia and in Asia in the Bering Sea and off Kamchatka. They spawn in lakes in the period July to October and spend 2 to 4 years in freshwater before going to sea for a further 3 years. The average weight at return is from 2.7 to 4 kilogrammes.

At sea they are a metallic greenish blue with black specks on the back with a lighter belly and a translucent green tail. In freshwater they become bright red with a silver green head. The deep red of the flesh is retained after cooking which makes them the premium fish for canning but they also obtain good prices in fresh and frozen form when they are sold whole, in steaks and in fillets. A relatively small number are smoked.

1.5 FISHING METHODS

Most salmon are still caught by commercial fisheries, using a number of different methods, each of which has implications for the quality and cost of the landed fish. It should also be noted (and this is discussed further in Chapter 3) that the types of fishing used in particular areas are very much constrained by national and international regulations relating to permitted fishing gear and fishing methods.

1.5.1   Trolling or Longlining
This method involves the use of lines and hooks, usually baited with fish or artificial lures, suspended from or towed by small fishing boats. Troll fishing takes place in North America and in the Greenland salmon fishery.   It is a relatively inefficient method of fishing since limited numbers of salmon can be taken at any one time and it may involve considerable effort in locating fish at sea.   However, because the fish that are caught are "ocean bright" and because the capture method causes little damage to the fish the catches command a premium price.   Trolling is used to catch the high value chinook and coho but, as indicated earlier, in recent years increasing quantities of pinks have also been taken by troll boats.   Overall, troll boats are responsible for a relatively small proportion of the total catch, taking for instance less than 10% of the total Alaskan catch.   Fish are gilled, gutted and iced aboard the boats and in some cases the boats also carry refrigeration facilities.   Catching is usually within one day's access to main landing processing facilities.

1.5.2   Drift and Gillnetting
A wall of net, typically 3 to 10 metres high, is suspended from a boat or marker and permitted to drift into the path of the fish which then become entangled in the mesh as they attempt to evade or pass through.   Since the gills of the salmon become entangled the fish are not able to escape.   The salmon are then picked from the net as it is reeled aboard.   As fish rapidly die and lose condition, nets must be promptly emptied. Monofilament nets are increasingly used for this and due to their light weight and small size often form the basis of illegal fisheries.   In the North Pacific, the boats used are small and do not go far from their base at which fish is unloaded daily.   Gillnet catches are large - for instance nearly 50% of the Alaskan catch is taken in this way.   It is also a more efficient fishing method than trolling and thus capable of putting greater

Figure 1.4: Fishing methods: troll, gillnets and
seine

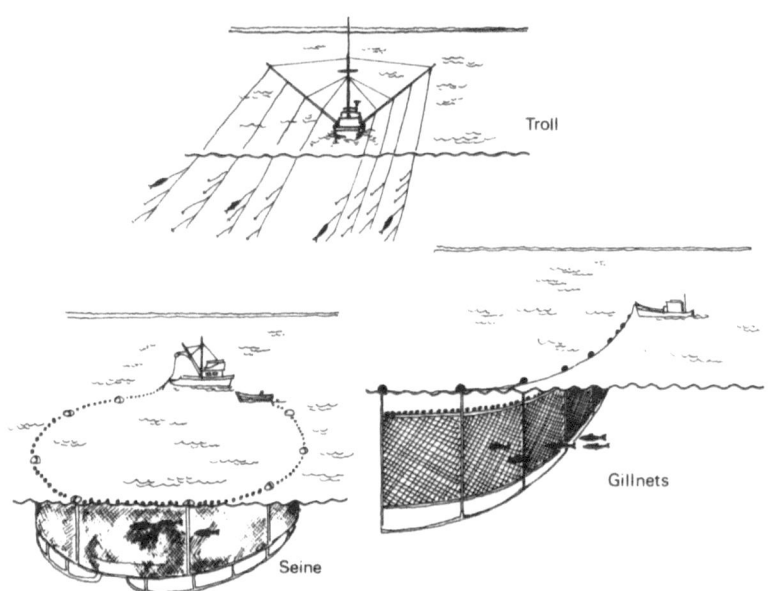

pressure on salmon stocks.

Since it is possible for the fish to become bruised, scaled and damaged in fishing and unloading it is a very skilled job to extract fish from the nets so that minimal damage results. Therefore, a lower average quality reduces price levels below that of troll caught fish.

1.5.3   Seine Netting
A floating net is set in a circle around shoaling fish and is pulled closed to make the catch. Purse seines are specifically designed to close around the bottom to prevent escapes downwards. In the Pacific salmon fishery, seine boats patrol the areas along beaches when the salmon are heading for their spawning rivers. After the net is laid the ropes are hauled in, frightening the fish which swim inwards where they are caught by the closing net. Because salmon are a schooling fish the catches can be heavy and up to 1500 fish have been taken in one net. Thus this method can also exert considerable pressure on stocks.

The fish are less likely to be damaged by the net than with gillnetting, but the pressure of the fish in the net and problems transferring fish to the hold may cause damage so that generally the quality of seine net caught salmon is considered to be below that of troll and gillnet salmon. Most salmon are also nearer spawning grounds by this time and are usually duller and softer in quality.

1.5.4   Land Based Fishing
Many different techniques are used, probably the most common being the use of set gillnets. Set gillnets work on the same principle as drift gill nets except that they are staked or anchored to the shore or inshore seabed in some way and normally handled by small boats or from the shoreline. Most of the catch in Japanese waters is taken in this way. Beach seines are used particularly in estuaries or lower river fisheries, the seine being set by small boat and hauled on to land. Bag nets or trap nets are also used, involving the use of systems of nets which direct the oncoming fish towards the trap net, so designed that it is difficult for the fish to escape. As most of the fish are held free swimming in the net the quality can be very

Figure 1.5: Land-based fishing methods

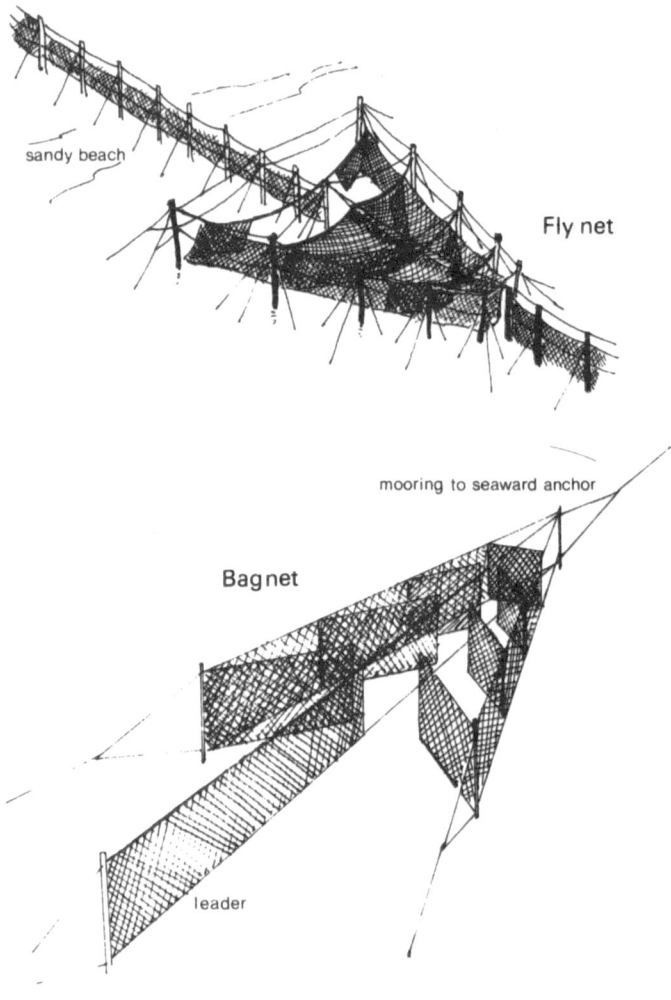

good. Hand held and operated nets are used in some areas, normally estuaries and rivers. In the Yukon there are fish wheels which are turned by the river current and which scoop the fish automatically up into baskets.

Finally as the fish approach the spawning grounds they may be taken by ladders leading into traps, although the use of such methods is closely regulated to allow adequate numbers of fish to escape to spawn. The quality of the fish is usually poorer by this stage and this method may be commercially less attractive.

1.6 ARTIFICIAL ENHANCEMENT AND RANCHING

Artificial enhancement involves stripping salmon of eggs and milt and fertilising the eggs artificially. The eggs are incubated and the young salmon reared in hatcheries until able to survive in the natural environment. Since the rates of survival of eggs and fry in the wild are low, artificial rearing of young salmon theoretically provides a means of increasing salmon stocks. The establishment of hatcheries has been widespread both for Atlantic and Pacific stocks. Many of these hatcheries have been established to enhance wild stocks for sport fishing or to compensate for river regulation or deterioration of water quality, but others have commercial objectives.

Thus public authorities have established hatcheries in order to enhance the resource base available for commercial fishermen. These artificially reared stocks are released into river systems to become part of the common property resource. This has occurred in both the Atlantic and the Pacific, although the enhancement activities in the Pacific have been the most significant. It has also been used in the Baltic where the salmon stock is now almost entirely dependent on hatchery production.

In recent years interest has increased in the use of privately run salmon hatcheries for salmon ranching with the objective of generating profits for the owners of the hatcheries rather than for fishermen. This is achieved by relying on the homing instinct of salmon which return to the hatchery from which they emerged, at which point they are captured by nets, traps or raceways, and slaughtered. The costs of harvesting the fish in

Figure 1.6: Salmon ranching

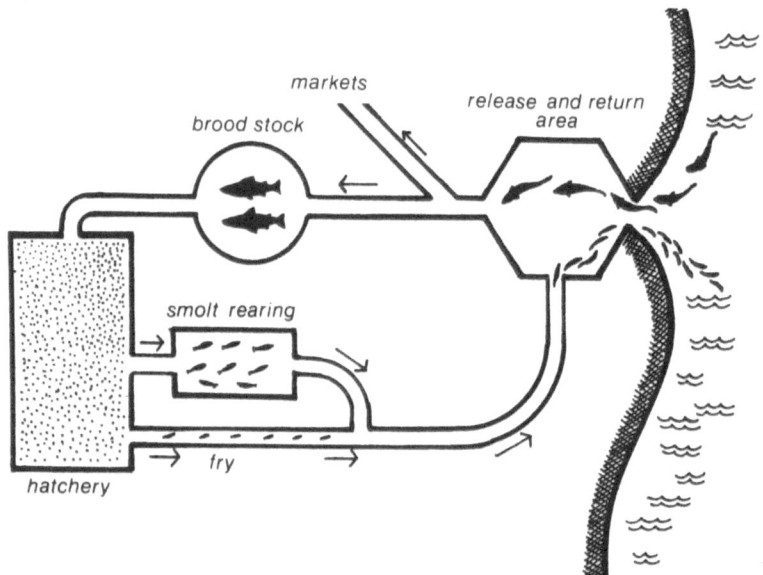

this way are low compared with fishing costs, but of course the cost of rearing the young salmon in hatcheries has to be offset against this. An intermediate operation is that in which fishermen subscribe or pay catch fees to support hatchery production in the river areas around which they fish.

These techniques are fundamental in the transplantation of stocks and the development of salmon fisheries in the Southern Hemisphere.

1.7 SALMON FARMING

A further development, particularly with Atlantic salmon and most markedly in Norway, has been the farming of salmon by retention in captivity throughout the life cycle rather than releasing the fish after the hatchery stage. The most usual system uses floating cages suspended in salt water in sheltered bays and estuaries. The cages are reached either by boat or by walkways from the land. Some salmon farms have also been established by closing off the arm of a fjord or of a bay and retaining the salmon within this enclosed area. There are also land based systems where

Figure 1.7:   Fish farming

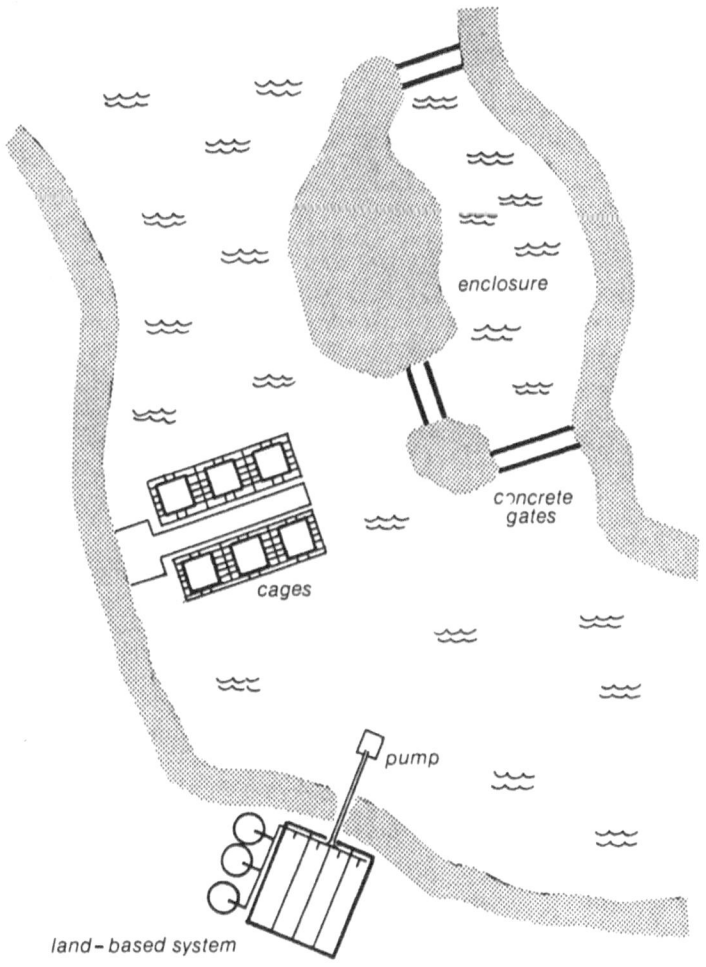

seawater is pumped from the sea into tanks or raceways. The salmon are fed on high protein diets normally based on fish meal.

For a number of reasons the earliest successful developments have involved Atlantic salmon. They grow quickly and can be reared to a marketable size in 2 years of on-growing, faster than with some of the other species. They sell at a relatively high price which is important since the heavy costs of feeding, maintaining the fish in captivity and occasional losses through disease have to be covered. Pacific salmon can however also be farmed and interest in the farming of coho has been growing recently with significant production now taking place in Japan, Chile and Canada. Chinook are also being grown successfully in New Zealand.

Salmon farming does not produce a cheap product, but with careful attention and handling at all stages it produces a very high quality product and one less subject to seasonal variations, which as a result has found rapid acceptance in world markets. Because of the high costs it seems likely that the main interest in farming will continue to be with high value species such as Atlantic salmon, chinook and coho rather than with the lower priced pinks and chums.

Salmon go through a number of different stages in the marketing channel on the way to the final consumer and may eventually end up in markets far distant from the point of landing. The types of processing and products are described below.

1.8 HANDLING AND PROCESSING

1.8.1 Fresh or "Wet" Salmon

The amount of processing involved is very small. Fresh salmon are marketed whole "in the round" or sold dressed, i.e. gutted and gilled. Sometimes the head is also removed and fish are sold in steaks as well. The fish are preserved in ice and are normally consumed within 3 days of slaughter if sold whole and 4 days if they have been gutted. They are normally packed in insulated boxes.

1.8.2 Frozen Salmon

Frozen salmon are usually headed and dressed before being individually blast frozen at minus

40°C or less for at least twelve hours. Fish may be vacuum packed before freezing; most are glazed and packed in polythene bags and in boxes after freezing. Frozen salmon are stored between -20 and -30°C and can be kept for at least a year, with reglazing after six months. Some reprocessing may take place before final sale.

1.8.3   Smoked Salmon

Fillets of salmon are first chilled in a brine solution or salted, frequently with added spices and flavours. The most common form of smoking is a cold smoke which involves smoking at less than 30°C. Hot smoking is also used and this cooks the product at temperatures of over 37°C resulting in a taste which is much stronger than that of cold smoked salmon. Smoked salmon is a fairly perishable product and has to be kept iced or chilled before consumption. Its shelf life in this form is similar to that of fresh iced fish. Fresh smoked salmon is sold loose wrapped, in vacuum packs and in controlled atmosphere packs. In addition smoked salmon can be sold in frozen form, usually in vacuum packaging with rigid outer cases. Whole sides of smoked salmon are sold but in recent years smaller packs of ready sliced smoked salmon have become increasingly popular.

1.8.4   Salmon Canning

The history of the early development of the Pacific salmon industry is a history of the development of salmon canning. Given the remoteness of landing points from markets it was then the best way of preserving the product and the first cannery was established in California in the 1860s. Nevertheless in spite of more recent developments in refrigeration and the preservation and transport of fresh fish, canned salmon remains very popular.

Salmon are canned in packs of 1/4 lb, 1/3 lb, 1 lb and 4 lb. Fish are delivered whole to the processing plant located near to landing points. The roe is extracted and the fish are headed, gutted, cut into pieces and put into cans which are then sealed. The sealed cans are placed in a retort for cooking and tested carefully for leakage. Initially the packs are shipped without labels and are known as "brights". The brand label of the seller is added later after the

19

salmon are sold by the canner. This may be either the brand label of the canners themselves or that of the wholesaler or retailer.

1.8.5   Salmon Roe
Roe are the unfertilised eggs of salmon. These are soaked in brine, packed in polythene lined wooden boxes, salted and cured at room temperature for several days. Only small quantities of the roe of Atlantic salmon are taken, but sales of Pacific roes particularly the roes of chum salmon, are considerable, especially in Japan.

1.8.6   Other Salmon Products
There are a number of other salmon products, often by-products of salmon smoking. The volumes involved in total are small but they make import-ant contributions to the economics of smoking processes. The main products involved here are pâtes and pastes. Marinated salmon is also sold, using a wide variety of different marinades although the most popular is the dill and mustard marinade used in gravlaks salmon.

With increasing consumer interest in pre-prepared products a number of prepared dishes have been marketed involving for example the addition of various sauces. The volumes sold at present are small but they are growing.

1.8.7   Processing Choices
A number of factors are involved in deciding which processing route is followed. Firstly, it depends on the species of salmon. Buyers and processors aim to allocate a particular species to the destination where the combination of species characteristics and market demand ·combine to yield the best price. Thus chinook, coho and Atlantic salmon go mainly to fresh and frozen markets or for smoking because their appearance, size and taste command a premium in these markets. Sock-eye which is another high value salmon is canned in large quantities but premium sockeye is increasingly being frozen rather than canned. The lower value pinks and chums are canned in larger proportions than other species. Salmon smokers have traditionally preferred larger fish and thus most chinooks and high proportions of the larger coho and Atlantic salmon are smoked. In

recent years however, increasing proportions of chums and occasionally pinks have been smoked in order to provide a section of the market with a lower priced smoked product. Salmon trout, mentioned earlier, competes with salmon here.

Choices are also influenced by the location of landing or harvesting in relation to processing facilities and markets. Normally fresh salmon commands a price premium over frozen salmon and freezing costs are avoided so, given accessible markets, fresh sales are preferred. Much of the Atlantic salmon produced in Europe can be delivered fresh to the adjacent European markets in this way. On the other hand, because of the remoteness of landing points much of the Alaskan catch is processed by canning or freezing in Alaska and then transported onwards to markets. Through time improvements in methods of preserving and handling fish have led to changes in these patterns. Improvements in refrigeration techniques and transportation methods have increased the proportion of fish which is frozen rather than canned. More recently developments in the storage and transportation of fresh salmon, particularly through air freighting, have widened markets geographically.

Often allocations of product between markets and forms of processing will change from year to year depending on the state of the market and the level of supplies. This is particularly the case for wild Pacific salmon where harvests change from year to year and processors have to endeavour, as well as meeting market needs, to schedule flows of product through their processing plants. Existing levels of stocks of canned and frozen salmon also influence choices. For instance, carryovers of large canned salmon stocks from a previous year may mean that more of next year's harvest is likely to be diverted to the fresh/frozen markets and vice versa.

FURTHER READING

Alaska Seafood Marketing Institute, Alaska Salmon
J E Bardach, J H Ryther, W O McLarney,
    Aquaculture: The Farming and Husbandry of
    Freshwater and Marine Organisms 1972 Wiley,
    New York
Department of Agriculture and Fisheries for
    Scotland, Cmnd 2096, Scottish Salmon and Trout
    Fisheries 1st Report 1963
S Drummond Sedgwick, The Salmon Handbook 1982
    Andre Deutsch, London
D Mills, Salmon and Trout: A resource, its
    ecology, conservation and management, Oliver
    and Boyd, Edinburgh
T J Pitcher and P J B Hart, Fisheries Ecology 1982
    Croom Helm, London
J Reardon (ed), Alaska's Salmon Fisheries, 1983,
    Alaska Geographic, Volume 10, no. 3
J E Thorpe (ed), Salmon Ranching 1980, Academic
    Press, London and New York

Chapter Two

MARKETS, SUPPLIES AND TRADING PATTERNS

2.1 INTRODUCTION

One of the fascinating characteristics of the international salmon business is the complexity of its trading patterns. The purpose of this chapter is to attempt to identify these main trading flows world-wide and to indicate how they have been changing through time. The analysis starts with harvesting patterns and this is followed by an investigation of the major markets for canned, fresh, frozen and smoked salmon. One cautionary point however must be noted at the outset. The quality of statistical data available from the United States and Canada is high but gaps remain even with this data, while statistical data relating to other countries ranges from less than adequate to virtually non-existent. The data available does make it possible to build up a broad picture of the structure of the international marketing system but figures throughout should be regarded as indicative rather than precise. There are also often considerable time lags before data is available so that although 1985 or 1984 data is used wherever possible, it is sometimes necessary to rely on data from earlier years.

2.2 SALMON SUPPLIES

2.2.1 Overview
Commercial landings of salmon since 1970 can be seen in table 2.1.
From this table it can be seen that in volume terms commercial landings are dominated by landings of Pacific salmon, particularly by the pink, chum and sockeye species and that these

Table 2.1:   World production of salmon by species, 1970-83, tonnes

	1970	1971	1972	1973	1974	1975	1976	1977	1978	1979	1980	1981	1982	1983
Atlantic														
wild	11277	10808	10901	12704	11873	12147	8692	8963	7166	8063	10107	9916	8562	7518
farmed	10	60	350	700	850	1200	1800	2537	3750	4620	5598	9555	13792	17795
TOTAL	11287	10868	11251	13404	12723	13347	10492	11500	10916	12683	15705	19471	22354	25313
Pacific (wild)														
Chinook	24000	25800	24000	27400	24100	24800	27000	26300	25400	25200	23100	21700	21700	19784
Chum	113700	105900	137700	125300	121100	127900	123200	118700	122800	150100	166100	179600	191300	188850
Coho	43300	43400	34200	38900	43000	31900	38600	31200	31200	34700	32800	28500	39900	34445
Pink	133500	178800	93900	151200	94000	170000	145400	224500	183900	249200	225700	281100	242800	252830
Sockeye	106400	78900	43900	55600	126400	38500	60100	64500	75600	109000	108300	133400	123100	164000
Masu	-	-	3300	4200	3100	3900	3800	3800	3600	2700	2700	3000	3000	3300
TOTAL	420900	432800	337000	402600	411700	397000	398100	469000	442500	570900	558700	647300	621800	663209
TOTAL ALL SALMON	432187	443668	348251	416004	424423	410347	408592	480500	453416	583583	574405	666771	644154	688522

Source: FAO Production Statistics (various)

landings have been rising over the past decade. Production of farmed, mainly Atlantic salmon, although less significant in volume has also been rising.

2.2.2   The North Pacific Fishery: Supplies
The Pacific salmon harvest is almost entirely of wild salmon with the largest landings being of pink and chum salmon (see figure 1.1). Production of farmed Pacific salmon, while growing, is currently only around 6000 tonnes.
Four nations are responsible for the bulk of Pacific salmon landings: the United States, Canada, Japan and the USSR.

Of these, the United States is the most significant accounting for approximately 45% of the total in 1983, followed by Japan with 25%, the Soviet Union with 16% and Canada with 14% (see figure 2.2), although the percentages fluctuate somewhat from year to year.

The United States catch is dominated by landings of sockeye and pinks and by landings from Alaska.   In 1985 92% of United States commercial landings by weight came from Alaska, the remainder coming from Washington State (7%), Oregon and California.   The substantial absolute increase in US landings over the period 1966 to 1985 (table 2.2) is almost entirely due to increased landings from Alaska.

Three factors account for this increase.   The first is the decreased Japanese high seas salmon catch due to the establishment of 200 mile zones by the United States in 1977.   This has reduced high seas interception fishing of Alaskan stocks and consequently greater numbers of salmon have been returning to Alaskan waters each year. Secondly, for some years climatic conditions are thought to have been particularly favourable to Alaskan stocks.   Finally, and of some concern for the future, Alaskan stocks are now more heavily exploited by the Alaskan fleet.   Landings from other US states by contrast have been static or declining, reflecting interception fishing, excessive exploitation of stocks and more recently, possible effects of unfavourable climatic conditions due to higher water temperatures and deteriorating environmental conditions in freshwater.   The latter have been blamed on urban developments and timber operations.   Landings did improve substantially

Figure 2.1:   Pacific salmon landings by species

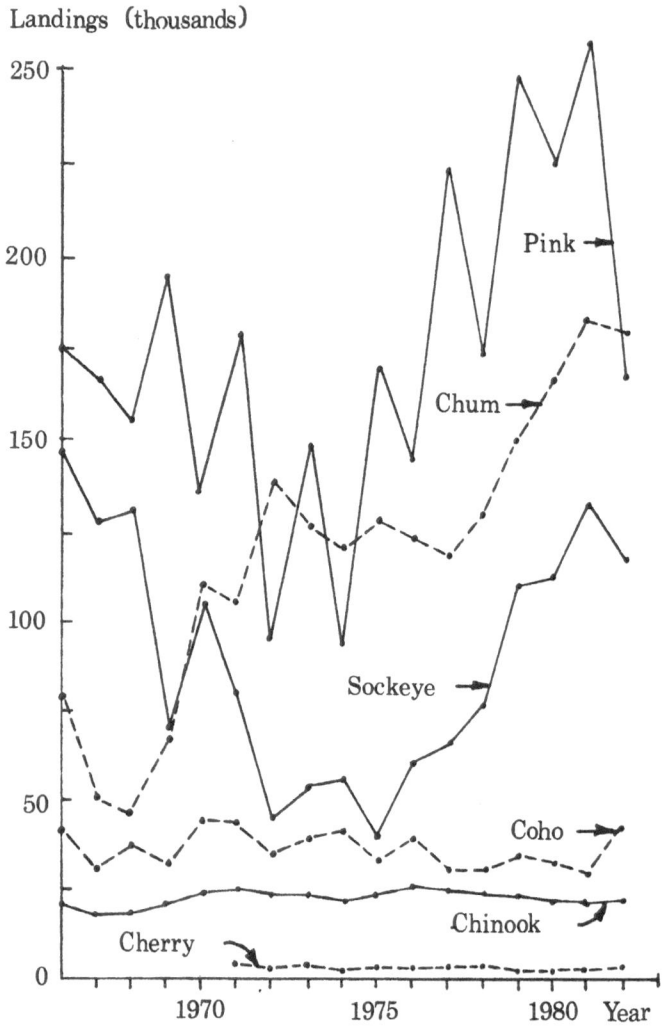

Landings (thousands)

Table 2.2: Salmon harvests in Pacific North America 1983-5

	Fish Millions	Weight Th. tonnes	Ex vessel Value $US	Ex vessel Average Price $US/Kg
United States				
Alaska				
1983	127	287.1	328	0.52
1984	132	298.6	338	0.51
1985	144	330.7	370	0.51
Washington				
1983	4.0	10.9	17.4	0.73
1984	3.8	12.2	14.4	0.60
1985	9.2	24.9	45.2	0.82
Oregon				
1983	0.4	0.8	0.30	1.67
1984	0.31	1.4	4.5	0.50
1985	0.62	2.6	8.8	1.54
California				
1983	0.29	1.1	4.1	1.75
1984	0.33	1.3	7.6	26.2
1985	0.8	2.1	11.7	2.54
Canada (British Columbia)				
1983	31.4	74.0	0.5	
1984	11.8	49.2	105	0.97
1985	42.0	83.6	148	0.80

Source: derived from data in Pacific Fishing

Figure 2.2:   Pacific salmon landings by country

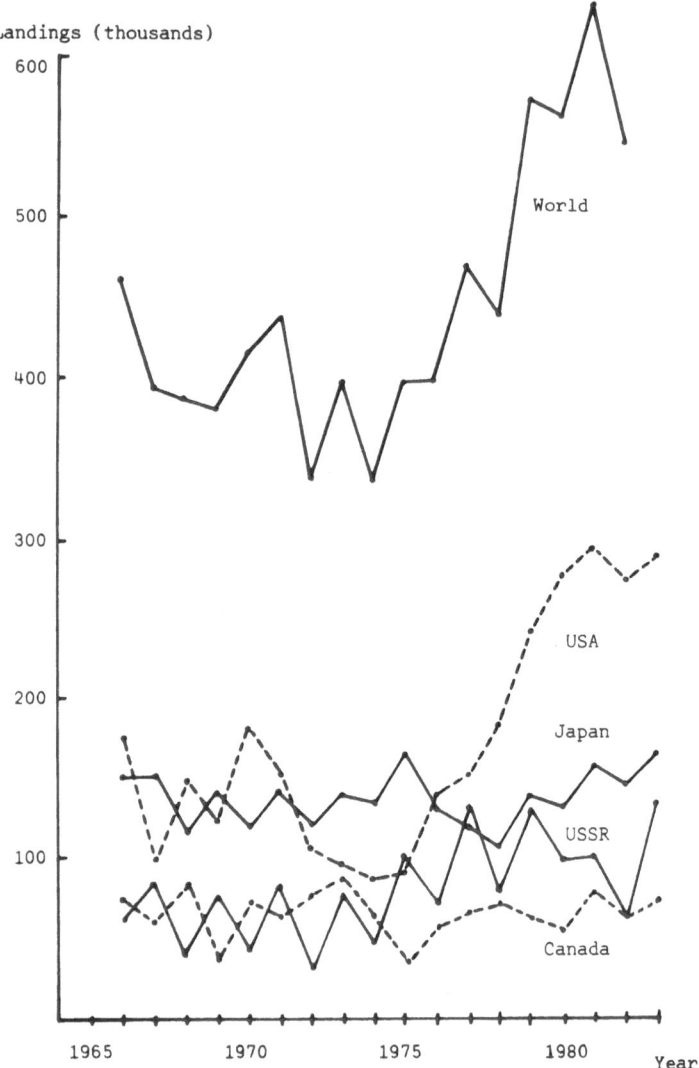

in Washington and British Columbia in 1985 but it is a little early to predict whether this is a reversal of the previous trend or merely a short term fluctuation in patterns.

Japanese salmon supplies have increased as well, doubling over the period 1974 to 1984. This has largely been due to an extremely successful hatchery programme for chum salmon. The Japanese Fisheries Association estimated that this source accounted for 45% of Japanese salmon landings in 1983 (1). Japan also has a high-seas salmon fleet which fishes under international agreement in Soviet waters. Its 1983 quota was 43,000 tonnes. Finally, the farming of coho is developing with a harvest of 4,400 tonnes in 1984. This is expected to grow further in the future (2).

Less is known about the Soviet catch except that it is dominated by landings of pink salmon. Like the Japanese catch, landings are increasingly from returns to expanding hatchery programmes.

Canadian landings have not enjoyed the same general expansion as that of other countries, in spite of a successful enhancement programme. This is probably due to heavier exploitation of fish on their return routes to Canadian rivers by interception fisheries in Alaska and elsewhere. Heavy fishing has also been taking place in Canadian waters. Sockeye landings are an exception to this and have grown considerably over the past decade as the result of a particularly successful enhancement programme.

In addition to commercial landings, there is a substantial sport fishery, particularly in North America. Thus it is estimated that in 1976 the marine sports fishery for chinook alone in North America amounted to 870,000 fish or approximately 23% of the total catch. For coho the comparable figure was 17% (3). Clearly for questions of overall management of stocks, these sports fisheries are of considerable importance. They are of considerable economic significance as well in terms of the multiplier effect on regional earnings through tourist revenues and are clearly politically significant in the formulation of management policy.

2.2.3   Atlantic Salmon: Supplies

A number of nations have fisheries for Atlantic salmon (table 2.3). Landings of wild Atlantic salmon have shown some tendency to decline over the last decade possibly due to localised over-exploitation of stocks, although worsening freshwater environmental conditions may also have had some impact.

Table 2.3: Estimated landings of wild Atlantic salmon by country 1983

	Tonnes	%
UK	1016	15
Ireland	1656	24
Norway	1530	22
Canada	1424	21
Faroes	678	10
Greenland	310	5
Other	211	3
TOTAL	6825	100

Source: ICES provisional

By contrast the production of Atlantic farmed salmon is growing rapidly and now dominates total supplies.   This has risen from a negligible proportion of the total in 1970 to 90% of the total in 1985.   (Table 2.4)

Table 2.4: Supplies of Atlantic Salmon 1983-6

	1983	1984	1985	1986*
Atlantic Salmon:				
Wild Total	7518*	7000*	7000*	7000*
Atlantic Salmon:				
Farmed Total	17795	26748	36950	57600
of which from				
Norway	14956	22196	28600	38000
Scotland	2539	3912	6900	10000
Ireland	300	340	650	1250
Faroes		300	600*	1000
Other European			100*	200
North America			100*	200
Atlantic Salmon				
Wild and Farmed				
Total	25313	33748	43950	64600

* = estimates
Sources: S A Shaw and J Rana, <u>Markets for Scottish Grown Salmon</u>, and <u>Federation Europeene de la Salmoniculture</u>

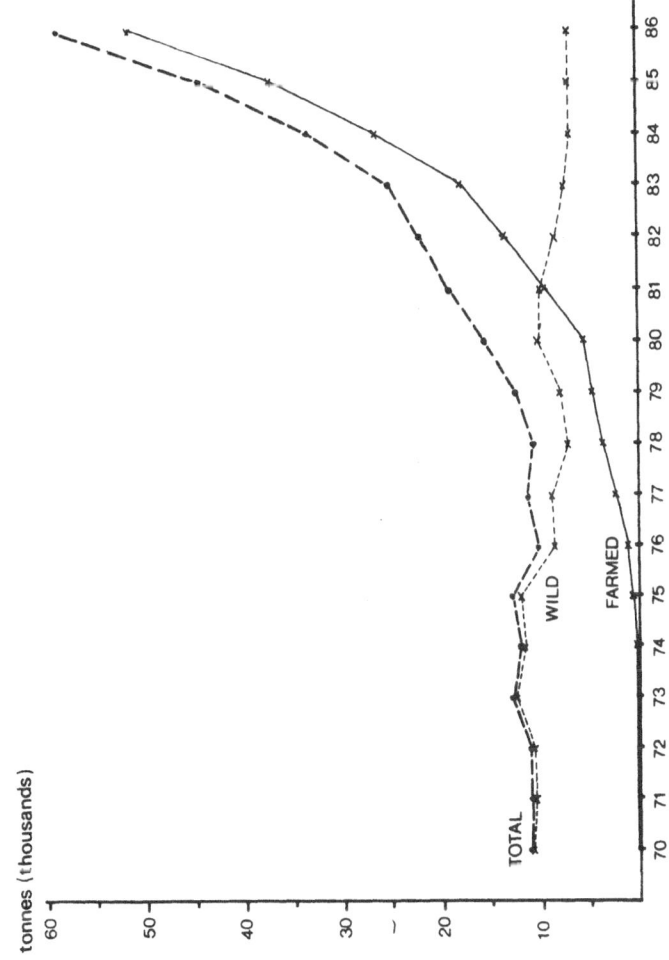

Figure 2.3: Supplies of Atlantic salmon 1970-86

As can be seen from table 2.4, Norway is the main producer and its production is expected to continue to expand but the farming of Atlantic salmon is now developing rapidly elsewhere as well, particularly in Scotland but also in Ireland, Iceland, the Faroes and Canada. Farmed Atlantic salmon represents only 3% of total world supplies but it has had a major impact in the higher quality section of the world market.

As with Pacific salmon, the Atlantic salmon sports fishery is of some significance. Thus in Newfoundland in 1980 sports fishermen caught 55,000 fish or 12% of the Canadian Atlantic salmon catch. In Scotland sports catches account for over 30% of the total recorded catch. Sports fishery figures are however excluded from the subsequent tables in this chapter which refer only to commercial landings of salmon.

2.3 DISPOSITION OF THE CATCH

2.3.1   Pacific Salmon
Where and in what form the salmon is marketed depends on many factors but in particular it depends on the species and the geographical relationship between the source of supplies and the major markets. Patterns of supply to the different markets also depend on the relative strength of demand in different markets at different times which is subject to change but despite this certain broad patterns are repeated from year to year and these are discussed in the following sections.

Looking first at North America, a noticeable feature is the low percentages of coho and chinook which are canned. This is shown in table 2.5 which gives illustrative figures for the United States. The pattern is similar for Canada, the other major producer of canned salmon.

Only small quantities of coho and chinook are canned because they command higher prices in fresh or frozen form. As a result, when they are canned it is likely to be because they are second grade, bruised or damaged fish. Coho and chinook are also bought in considerable quantities by salmon smokers in Europe and the United States. Chum catches are used in a similar way with a mixture of sales in fresh/frozen and in canned form. Pinks on the other hand are smaller than other species and are less sought after by the

Table 2.5:   Disposition of the US salmon catch by
             species 1984

Live product weight or its equivalent (estimated)

Species	Landings tonnes	Canned tonnes	%	Residual (fresh, frozen, tonnes	smoked) %
Chinook	9,848	295	3	9,553	97
Chum	51,411	12,853	25	38,558	75
Coho	23,168	2,310	10	20,858	90
Pinks	125,018	83,700	67	41,318	33
sockeye	104,176	31,992	31	72,184	69
TOTAL	313,621	131,150	42	182,471	58

Source: US Department of Commerce, Fisheries of
        the United States Pacific fishing 1985

fresh market unless from troll fisheries. They
make very good canned fish because they preserve
their colour through the canning process with the
consequence that the proportions canned are higher
than for some other species.   The final species,
Sockeye, is the premium salmon for canning because
of its deep red oily meat but it is also much
sought after in fresh/frozen form particularly in
Japan.   As a result, as the market for frozen
sockeye has developed in Japan, the proportions of
sockeye going for canning have declined.   Once
again, the normal pattern is mainly for first
grade fish to go to the fresh/frozen Japanese
market and for lower qualities to be canned.
     The absolute volume of salmon which is canned
has remained roughly constant in recent years but
the proportion of total landings going for canning
has been declining.   This can be seen in table
2.6 which shows the disposition of the Alaskan
catch.   Because demand for fresh and frozen
salmon has grown more rapidly than the demand for
canned salmon and the fresh/frozen market has been
prepared to pay higher prices, the bulk of the
increase in salmon supplies has gone to this
market rather than to the canners.
     The declines in the proportions of sockeye and
chum being canned are the most noticeable,
reflecting in particular the strong demand for
frozen sockeye and chum in Japan which has already
been mentioned.

Table 2.6: Alaska: disposition of landings 1971-83

	TOTAL Canned Pack 000 cases	SOCKEYE Canned Pack as % landings	PINKS Canned Pack as % landings	CHUM Canned Pack as % landings	COHO Canned Pack as % landings	CHINOOK Canned Pack as % landings
1971	2899	91	98	81	58	18
1972	1705	93	95	72	31	14
1973	1210	73	94	51	21	6
1974	1343	88	90	53	31	11
1975	1384	79	94	41	5	8
1976	2470	82	92	62	21	22
1977	2595	69	92	60	13	6
1978	2922	57	89	47	14	7
1979	3049	43	83	40	13	6
1980	4132	45	84	48	22	10
1981	4432	41	78	51	11	3
1982	2435	12	65	17	6	1
1983	3601	33	78	24	11	4

. Source : Seafood Business Report

In Japan itself, most of the Japanese catch is consumed by the home market. Here, most of the fish are sold salted and only small quantities are canned. The Japanese also consume considerable quantities of cured salmon roe. As will be seen in Section 2.5 in spite of large domestic landings, Japan is also the world's largest importer of frozen salmon, the latter being mainly imported from North America.

Little is known about the disposition of the Soviet catch since most of it is consumed domestically. The only Soviet salmon traded in appreciable quantities internationally is canned pink salmon which is exported in significant quantities.

2.3.2   Atlantic Salmon
Disposition patterns for Atlantic salmon reflect its premium price and quality position in international markets. Most Atlantic salmon is consumed fresh or is smoked. Relatively small quantities are frozen or canned at present since fresh Atlantic salmon command a premium price over frozen Atlantic salmon or canned salmon. Freezing costs are also avoided if the salmon are sold fresh. In most cases fresh sales are feasible because transport systems have been developed to get the product fresh to markets and because much of the production is now from fish farms which are less remote from their markets than many landing sites for Pacific salmon. Exceptions are mainly those quantities frozen during periods of glut or frozen for smokers, some of whom prefer to buy frozen rather than fresh salmon. Because the product is normally supplied fresh it has to be transported rapidly to markets. This requires either that the markets are close (Norwegian and Scottish supplies to European markets for example) or that fast but expensive methods such as air freight are used. This is discussed further in subsequent chapters but as an example the disposition pattern for Scottish farmed salmon can be seen in figure 2.4.

2.4   THE MARKETS FOR CANNED SALMON

The canned salmon business is dominated by a small number of countries: as already indicated the major exporters are the United States and Canada. The major consumers are the United

Figure 2.4: Distribution flows for Scottish farmed salmon

Tonnes 1984 (estimated)

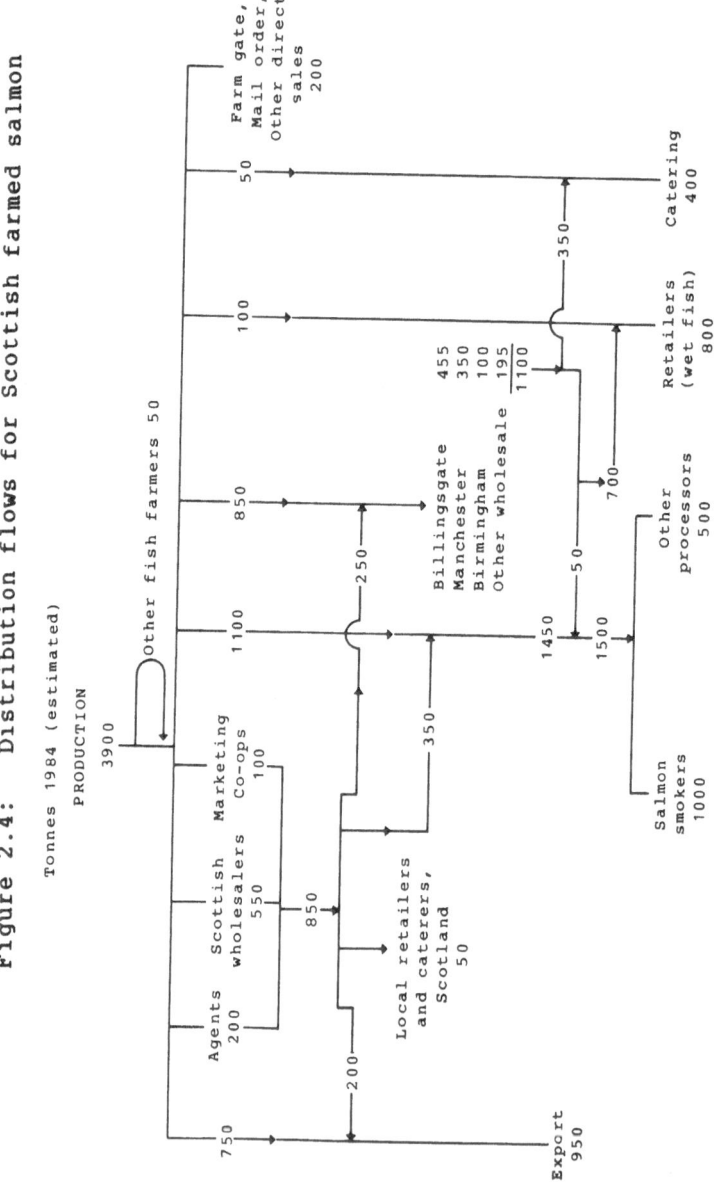

Kingdom, Canada and the United States, both of the latter consuming a large amount of their own production. Other substantial but smaller canned salmon markets exist in a number of other European countries, most notably the Netherlands, France, Belgium and Italy. There is a large market for canned salmon in Australia and canned salmon is exported to most of the countries of the British Commonwealth (table 2.7).

A characteristic of many canned salmon markets is that while total consumption increased in response to increased availability and lower real prices in the later 1970s, consumption per head has risen very little in the 1980s, as illustrated for the United States and the United Kingdom in tables 2.8 and 2.9.

Canned salmon markets are price sensitive (see Chapter 8) and exports are therefore affected by prevailing exchange rates. Thus exports to Europe were high in 1980 and 1981 because of declines in the US and Canadian dollars. When the reverse happens, either less is imported and consumed or the markets turn to other sources of supply such as the USSR. Upward movement in the dollar, given the importance of European markets to North American canners, in turn puts pressure on canners' profit margins in North America as European importers attempt to hold real prices down.

2.5 THE MARKETS FOR FRESH, FROZEN AND SMOKED SALMON

2.5.1   The Major Markets
Table 2.10 shows the major markets for fresh and frozen salmon and their sources of supply. These figures also include substantial quantities of salmon destined for smoking, especially of the higher valued Atlantic salmon, coho and chinook. The figures therefore include sales to smokers in the country of origin as well as substantial sales of both Atlantic and Pacific salmon to salmon smokers in the United Kingdom, Denmark, West Germany, France and other European countries. Thus the table should be regarded as an indication of the first destination of the salmon and of the form in which international trading takes place, rather than as indicative of final sales form. As an example however it has been estimated that 60-70% of salmon in European countries goes for salmon smoking.

Table 2.7: The major markets for canned salmon and their sources of supply (estimated) tonnes actual product weight 1984

Market	(1) Imports From Japan	United States	Canada	USSR	Other	Total	(2) Domestic Production	(3) Exports & Re-Exports	(4) Domestic Consumption	(5) Annual Per Capita Consumption kg
Japan	-	20	47	137	-	208	n.a.	655	n.a.	n.a.
United States	-	-	50	-	-	50	89,268	25,464	63,804	0.28
Canada	-	3,857	-	-	3	3,860	15,350	13,548	5,662	0.23
USSR	n.a.						n.a.	6,500	n.a.	n.a.
United Kingdom	155	12,838	5,534	2,667	274	21,468	-	478	20,990	0.38
Belgium & Luxembourg		875	1,008	612	1,076	3,571	-	623	2,948	0.29
France		291	542	1,527	779	3,139	-	6	3,133	0.06
Italy		123	267	75	71	536	-	-	536	0.003
Netherlands	275	1,853	271	1,196	2,124	5,719	-	1,744	3,975	0.28
Ireland	35	527	471	-	95	1,128	-	2	1,126	0.32
Australia	-	4,595	2,916	-	-	7,511	-	-	7,511	0.49
New Zealand	-	85	1,782	-	-	1,867	-	-	1,867	0.57
Others	190	400	660	286	-	1,536	-	-	-	-
TOTAL	655	25,464	13,548	6,500	4,422	50,593	-			

Source: Seafood Business Report, Trade figures and industry estimates

Notes:
1   Import statistics are used where possible in preference to export statistics.   There are often substantial differences between the two, however, reflecting differences in the timing of documentation and misclassifications.
2   Inventory adjustments over the year are ignored.   However in some years inventory changes have been substantial.
3   European export figures mainly represent re-exports usually to other European countries.

Table 2.8: US supply and consumption of canned
salmon 1970-84, tonnes actual product
weight

	Canned pack	Imports	Exports	Net Domestic Supply	Consumption per capita kg
1970	83,220	1,107	7,625	76,702	0.32
1971	76,409	704	6,270	68,843	0.32
1972	42,120	5,283	9,712	37,691	0.32
1973	32,558	3,565	7,695	28,428	0.18
1974	39,822	3,88	3,775	39,927	0.14
1975	35,420	1,481	10,232	26,669	0.14
1976	57,884	1,144	8,990	50,038	0.14
1977	68,413	266	9,655	59,024	0.23
1978	74,498	147	14,763	59,882	0.27
1979	70,234	197	23,006	47,425	0.23
1980	91,483	76	33,569	57,990	0.23
1981	98,745	32	28,801	69,976	0.23
1982	54,584	3	18,668	35,919	0.15
1983	79,297	120	24,716	54,581	0.23
1984	89,268	240	25,464	63,804	0.28 est.

Sources: DPRA: Alaska Salmon, Projected 1982 Market Conditions
US fisheries (various).

Table 2.9: Imports and consumption of canned salmon in the UK, 1970-84, tonnes actual product weight

| | Imports % from | | | | | |
	Canada	Japan	US	USSR	TOTAL	Per Capita Consumption kg
1970	11	64	15	10	25,400	0.46
1971	19	57	16	8	30,100	0.54
1972	23	49	24	4	30,600	0.55
1973	41	30	25	4	24,100	0.26
1974	37	42	16	5	14,900	0.37
1975	18	45	30	7	20,700	0.36
1976	21	47	23	9	18,200	0.32
1977	-	na	na	na	na	na
1978	39	19	34	8	13,900	0.24
1979	49	na	36	15	15,100	0.26
1980	30	3	62	5	24,800	0.44
1981	34	2	53	11	29,500	0.53
1982	41	2	48	9	16,200	0.29
1983	39	-	55	6	25,730	0.43
1984	26	1	60	12	21,468	0.38

Source : UK trade statistics

Table 2.10: The major markets for fresh and frozen salmon and their sources of supply, tonnes actual product weight

Major Markets	(1) Imports from Japan	United States	Canada	United Kingdom	Norway	other	Total	(2) Domestic Production	Domestic Exports & Re-Exports	(3) Total Domestic Consumption (1)+(2)-(3)
Japan		80,276	5,179	289	7,486		93,230	180,000	1,256	271,974
United States	109		4,813	104	3,968	575	9,569	182,471	103,889	88,151
Canada		9,814			11		9,825	22,771	22,123	10,473
United Kingdom	24	2,816	1,052		1,804	875	6,571	4,912	2,337	9,146
France		9,494	6,293	1,329	4,944	1,384	23,444		497	22,947
Sweden		2,409	1,720		1,015	374	5,518		520	4,998
Denmark	112	1,374	1,327	16	2,130	1,460	6,419		2,136	4,283
West Germany		840	749	59	3,059	788	5,503		120	5,383
Belgium & Luxembourg		1,029	479	125	600	634	3,067		290	2,777
Italy		239	511	7	83		840			840
Others	1,011		704		n.a.		n.a.	n.a.	n.a.	n.a.
TOTAL	1,256	103,889	22,123	2,337	16,016					

Notes:

1. These figures represent actual fish weight at each point. Note however that since frozen fish are headed and gutted after landing and since trade in salmon is dominated by the frozen trade the trade figures understate the round weight of fish involved by up to 20%.

2. Most fish from Norway and the UK is traded in fresh rather than frozen form. Some salmon is exported from the US and Canada in fresh form (particular to each other) but trade is dominated by frozen salmon.

3. These figures include re-exports. Denmark is particularly important as an importer and on-ward re-exporter to other European countries.

4. Import statistics are used where possible instead of export statistics due to data inconsistencies.

The largest consumer of fresh and frozen salmon in both absolute and per capita terms is Japan where salmon occupies a market position which is the envy of salmon dealers in other countries. It is a popular item in the diet across all income groups with particularly high levels of consumption at holiday times. It is worth noting that consumption prior to 1960 was minimal except in the northern coastal areas so that these high levels of consumption are of fairly recent origin. In North America and in Europe which are the other major market areas, per capita consumption is much lower but it has been growing (see table 2.11). In the United Kingdom consumption has risen by over 30% since 1978.

Table 2.11: Fresh/frozen salmon consumption trends, Japan, France and the UK, 1978-83[1] actual product weight

	JAPAN[2]		FRANCE		UNITED KINGDOM	
	tonnes	kg per capita	tonnes	kg per capita	tonnes	kg per capita
1978	154,000	1.34	12,824	0.24	7,005	0.12
1979	193,100	1.67	15,156	0.28	7,861	0.14
1980	171,700	1.47	15,674	0.28	7,155	0.13
1981	233,100	1.98	16,721	0.31	9,170	0.16
1982	263,200	2.22	18,033	0.33	9,386	0.17
1983	273,600	2.29	20,825	0.38	10,253	0.18
1984	271,974	2.25	22,947	0.41	9,146	0.16

Sources: NMFS
          DAFS
          Trade Statistics
          Industry Estimates
1   includes salmon which is subsequently smoked and in some cases re-exported
2   includes relatively small quantities of canned salmon

2.5.2   Species Preferences

There are distinct differences in the preferences for various types of fresh or frozen salmon in the different markets. These are indicated in figure 2.5.

These different preferences relate to differences in tastes, appearance, fat content, etc. Sensitivity to price also varies between countries and affects choice of species. For the interested reader the factors which affect demand are discussed more fully in Chapter 8.

2.5.3 The Impact of Farmed Atlantic Salmon

One of the most notable recent changes in international salmon markets has been the impact made by the rapidly growing production of Atlantic farmed salmon. The salmon is of high quality, reared and harvested under carefully controlled conditions and sold at the top end of the price range for salmon. Initially in the early and mid-1970s farmed salmon was mainly sold in Europe, especially to be consumed fresh. Subsequently sales to salmon smokers in Europe have also developed to significant levels. The largest of the markets in Europe are in France, West Germany and the United Kingdom. These European markets are continuing to grow but more recently other additional markets have been developed. A substantial market has been opened up in the United States for fresh salmon air-freighted from Europe (see table 2.12). In turn this success

Table 2.12: Norwegian exports of fresh farmed
        salmon 1981-4

	Weight (tonnes)			
	1981	1982	1983	1984
France	1758	2300	3263	3490
West Germany	1066	1527	2028	2431
Denmark	1058	1149	1940	2141
Sweden	533	600	794	879
U.K.	338	507	855	1183
Switzerland	266	328	506	516
Belgium/Lux	247	300	484	653
Spain	110	257	359	751
Netherlands	63	97	177	280
U.S.A.	8	711	2405	4640
Japan				183
Others	60	92	203	146
TOTAL	5507	7868	13014	17293
Annual % change	n.a.	43	65	33

Figure 2.5: Some characteristics of the major
markets for fresh and frozen salmon

Market	Species Preferences	Main Uses	Further Comments
Japan	Chum large consumption least expensive and most abundant.	Salted or smoked and sliced.	Mainly from Japanese landings of enhanced stock. Very price sensitive market.
	Sockeye	Salted and sliced.	Large volume imports from USA in frozen form.
	Pinks	Sliced and salted.	From domestic landings and imports.
	Coho	Sliced and salted.	Imports from USA.
	Chinook	Non-salted for use in Western-type restaurants.	Limited but growing market. Red Kings only.
	Atlantic	Small but growing volumes air freighted for raw consumption.	Limited high price market.
United States	All species consumed	Chinook and Atlantic mainly used for smoking and white table-cloth restaurants. Coho also used for smoking and fresh consumption Chum) mainly) eaten Pinks) fresh.	Big differences in consumption in different States with consumption greatest in Coastal states. Differences in regional preferences by species.
France	Large market for Coho and Atlantic salmon but substantial quantities of chinook, Pinks and Chum also imported.	Estimated that 50% of fresh Atlantic salmon and 70% of imported Pacific salmon is smoked. Prefers fresh rather than frozen fish for non-smoked consumption	Species conscious and well informed market

Market	Species Preferences	Main Uses	Further Comments
United Kingdom	Consumes own wild and farmed Atlantic salmon and imports Atlantic salmon from Norway. Substantial importer of frozen Chum and Coho.	Estimated that 75% of Atlantic salmon smoked. Most of non-smoked salmon consumed is fresh Atlantic salmon. Most imported frozen salmon is for smoking.	Significant exporter of smoked salmon. Little distinction between Pacific species at consumer level. Very price sensitive market for Pacific salmon.
West Germany	Half of consumption now of Atlantic salmon. Also imports substantial quantities of other Pacific species.	50-70% of imports for smoking. Preference for Atlantic salmon for non-smoked consumption.	Varying regional preferences. Imports substantial quantities of smoked salmon. Most of local smoked salmon consumed in Germany and not exported.
Denmark	Imports substantial quantities of Atlantic salmon and Chum.	Used for fresh consumption and for smoking.	Important for entrepot trade as importer and re-exporter of fresh and frozen salmon and exporter of Atlantic and Pacific salmon smoked in Denmark.

45

has prompted considerable interest in the farming
of Atlantic salmon in the United States and Canada
where major developments in the aquaculture of
both Atlantic and Pacific salmon are now underway.

2.6 SMOKED SALMON

It is unfortunately not possible to identify with
any precision the volumes of salmon which are
smoked as these categories do not feature in the
main trade or catch statistics. Estimates from
industry sources suggest that the size of the
major markets for smoked salmon are approximately
as shown in table 2.13, but the range of error
associated with these figures is likely to be
considerable.

Table 2.13: Smoked salmon: the major markets 1984

(estimated)
tonnes actual product weight

	Imports	Production	Exports	Consumption
France	743	9,000	553	9,190
United Kingdom	68	4,500	594	3,974
United States	329	10,000	--	10,329
West Germany	1,301	4,000	120	5,181

-- small
Source: Trade Statistics
Industry Estimates

France, the United Kingdom, the United States
and West Germany are the main consumers, producers
and the main exporters (see table 2.14).   Denmark
is also a substantial producer but mainly for
export markets outside Denmark.
Exports from all those sources are destined
mainly for European countries and North America
but the total number of countries who import
smoked salmon, albeit in small quantities, is
quite large.   For example, in 1984 smoked salmon
was exported from the United Kingdom to 55
countries outside the European Community as well
as to most countries inside the European

Table 2.14: International trade in smoked salmon 1984 . (estimated)

(tonnes)

| Importers | Exporters | | | | | | | | | TOTAL |
	United States	Canada	United Kingdom	Ireland	Denmark	France	Norway	Netherlands	Other	IMPORTS
United States		44	130	28	6	22	71	3	25	329
Canada	119		143	--		20		33	23	338
U.K.	--	16		7	6			--	39	68
Ireland	--	--	18	--				--	2	10
Denmark	--	--	--	--				--	15	15
France	53	--	125	50	285		40	--	220	743
Belgium and Luxembourg	--	--	25	17	121	71	2	5	20	261
West Germany	--	--	--	47	1033	13	72	46	90	1301
Italy	--	1	40	--	196	236	--	24	--	497
Australia	--	34	38	--	135	--	--	--	--	207
Sweden	--	--	--	--	42	--	45	--	--	87
Switzerland	--	--	22	--	196	--	21	--	--	239
Other	20	107	63	8	330	191	29	5	--	753
TOTAL EXPORTS	192	202	594	157	2320	553	280	116	434	4848

-- small

* estimates

Source: Trade Statistics

47

Community. The largest markets outside Europe and North America are those in Hong Kong, Australia and South Africa. Unlike its fresh/frozen market, the Japanese market for smoked salmon is currently limited to several hundred tonnes only but it is expected to grow fairly rapidly both from its own production and through a growth of imports.

This pattern of trade with a large number of countries consuming only small quantities per head reflects the current status of smoked salmon as a high priced luxury product going to narrow markets. Successful attempts are however being made to widen smoked salmon consumption, in particular through the development of small pre-packed sliced smoked salmon products.

2.7 SALMON ROE AND OTHER PRODUCTS

The only other product marketed in measurable volume is salmon roe. Japan represents by far the largest market for salmon roes, consuming roes from her own supplies and importing from the United States, Canada and Norway. In addition to domestic production of an estimated 5000 tonnes in 1984, Japan imported 8544 tonnes from the United States, 629 tonnes from Canada and 31 tonnes from Norway. The roes come mainly from pink and chum salmon. Small quantities are also consumed in Europe.

No estimates are available of the size of the market for other salmon products but industry sources in Europe and North America report rapid increases in demand for terrines and pâtes, pre-prepared dishes, cutlets, etc.

2.8 SUMMARY

Nearly 700,000 tonnes of salmon are landed annually. About 600,000 tonnes of this are of Pacific salmon from the Pacific Ocean with the remainder Atlantic salmon from Atlantic waters. Most Pacific salmon is the harvest of fisheries since the production of farmed Pacific salmon, while growing, is small. Over 90% of the 65,000 tonnes of Atlantic salmon available in 1985 was farmed salmon, with Norway the major producer.

The trading patterns for salmon are complex. Large quantities of Pacific salmon are canned, the main markets for which are in the United States, the United Kingdom, Canada and Australia. Very little Atlantic salmon is canned. The other major salmon products are fresh, frozen, salted and smoked salmon. The largest markets for fresh, frozen and salted salmon are Japan, the United States, France, Canada and the United Kingdom. Substantial quantities of both Atlantic and Pacific salmon are exported fresh or frozen for smoking in the country of destination. The major producers of and markets for smoked salmon are in North America and Europe.

NOTES

1. Fisheries of Japan 1982 and 1983, Japan
    Fisheries Association.
2. Fish Farming International, December 1984,
    Volume 11, No. 12.
3. R A Fredin, Trends in North Pacific Salmon
    Fisheries in (eds) W J McNeil and
    D C Hinsworth, Salmonid Ecosystems of the
    North Pacific, Oregon State University Press
    1980, Corvallis.

FURTHER READING

Alaska Fisherman's Journal
W M Carter, Atlantic Salmon Newsletter, quoted in
    Fishing News International, December 1983
Department of Agriculture and Fisheries for
    Scotland, Salmon and Trout Farming in
    Scotland, Annual Surveys
Development Planning and Research Associates,
    Inc., Alaska Salmon Projected 1982 Market
    Conditions 1982 for Office of Commercial
    Fisheries Development, Department of Commerce
    and Economic Development, State of Alaska
Fish Farmer, Business Press International,
    published 6 times per annum, London
Fish Farming International, AGB Heighway published
    monthly, London
The Fishermen's News, Seattle, bi-monthly
Fishing News International, AGB Heighway published
    monthly, London
Fiskeoppdretternes Salgslag (Norwegian Fish
    Farmers Sales Association) Arberetning og
    Regnskap 1983, Trondheim
Food and Agriculture Organisation of the United
    Nations, Yearbooks of Fishery Statistics, Rome
Infofish Marketing Digest, Infofish Marketing and
    Advisory Services for Fish Products in the
    Asia/Pacific Region, Food and Agriculture
    Organisation of the United Nations, published
    bi-monthly, Kuala Lumpur
International Council for the Exploration of the
    Sea, Annual Reported Nominal Catches of
    Atlantic Salmon
J F Muir, R G Poulter, S A Shaw, R I G Morgan:
    Salmon Ranching in the Falkland Islands: A
    Feasibility Study. Unpublished Report for
    the Falkland Islands Government commissioned
    by the Overseas Development Administration,
    British Government 1982, London
Pacific Fishing, Pacific Fishing Partnership,
    published monthly, Seattle
Province of British Columbia, Marine Resources
    Branch, Fisheries Production Statistics of
    British Columbia, published annually
Seafood Business Report, Journal Publications,
    published quarterly, Seattle
Seafood International, AGB Heighway Ltd.,
    published monthly, London
United States Department of Commerce, National
    Oceanic and Atmospheric Administration,
    Fisheries of the United States, published
    annually, Washington

Chapter Three

THE MANAGEMENT OF THE WILD RESOURCE

3.1 INTRODUCTION

The focus now moves to the factors affecting the
availability of wild salmon. This chapter discusses the management of fishing effort while
Chapter 4 is concerned with salmon ranching and
the artificial enhancement of salmon stocks. As
the previous chapter has already suggested, there
are substantial differences between fisheries.
Japan has been steadily increasing salmon landings, a trend associated with successful private
stock enhancement programmes. In the North
American fisheries, experiences vary by area,
species and stock. In general landings have
risen but distribution has changed. Thus
increasing quantities of salmon are taken in
Alaskan waters partly at the expense of landings
elsewhere in North America, partly because of the
protection provided by 200-mile zones and partly
because of changing environmental conditions. In
the Atlantic, catches are currently considerably
below earlier levels. However in spite of this
diversity of experience, the same management
issues affect all salmon fisheries and are
inherent in the management of salmon as a renew-
able natural resource.
      The main objectives to be achieved in the
management of public salmon fisheries are first
discussed, followed by an examination of the
effectiveness of the various management tools
available. Some alternative schemes involving
stronger private rights are then examined.
Finally, the international dimensions of management issues are considered.

3.2 THE COMMON RESOURCE IN PUBLIC FISHERIES

Public fisheries are defined as those where fishermen harvest the same stocks in common waters. These therefore include the Pacific and Atlantic fisheries of the United States and Canada as well as some of the European salmon fisheries.

Many objectives have been suggested for the management of these fisheries but there would probably be general agreement, at least in principle, on the importance of two of them. The first is that fisheries should be managed so that fishing effort does not deplete the resource and as a result impair the prospects of exploiting the fishery again in the future. Secondly, in ideal circumstances the stock should be exploited at an economically efficient level. This means that the economic benefits or "rent" from the fishery should be maximised. Two conditions are necessary for this. First the difference should be maximised between the costs of exploiting the resource and the revenue derived from it to maximise benefits. Secondly, the inputs to achieve that level of catches should be combined in the most efficient, i.e. least cost, way so that welfare is not lost through wasteful use of resources.

A characteristic of salmon fisheries, however, which they have in common with other fisheries, is that if the resource is a public one, the fishery will not regulate itself to ensure that these objectives are achieved. Hence the main issues in management revolve around the use of government and inter-governmental powers to regulate the fishery. The history of government interventions in salmon fisheries is one of a continuous search for tools which will enable them to achieve these objectives effectively and cheaply.

The search for effective tools in managing the salmon resource is particularly difficult because it is a non-static resource ranging freely for considerable distances in the oceans. This makes it difficult to assess the implications of the controls, since these will change with changing stock and environmental conditions. It also makes it difficult to respond rapidly to changing conditions because of the limited understanding of complex interrelationships. As a result fishermen often lack confidence in the judgement of the controllers so that the necessary consensus for successful legislation is missing.

The problem, well illustrated in the burgeoning literature of fishery economics (1), can be stated simply as follows. Firstly, imagine a privately owned salmon trap which is the sole means used to catch salmon as they return to spawn in their native rivers. The owners have sole and permanent rights to the salmon which return to the trap but a decision has to be taken as to the numbers of salmon to be taken each season. If too many salmon are taken in one year, this will have an adverse effect on spawning and the numbers of returning fish in subsequent years. If the owners put a high positive value on the fish which return in future years then they will have an incentive to limit the numbers of salmon taken in any one year in order to conserve the resource for future years. If they fail to do this, the trap will not yield them an income in the future (2).
Most salmon fisheries, however, do not conform to this model, hence the management problem. The most common pattern is that of a large number of individual fishing boats exploiting a common resource which they do not individually own.
This is the case in North American and North Atlantic waters. Even in areas where private rights exist, such as the Scottish salmon fishery and the very few traps still allowed in North America, the right to fish in particular positions involves sharing the same returning salmon stock with other private rights on the same river or coastal area.
The open access to the common resource creates an inherent tendency to over-exploit stocks. No single fisherman is motivated to conserve the resource since even if they individually exercise restraint, the salmon are still likely to be taken by others not behaving in the same way. The more fishermen there are and the more geographically dispersed the resource, the more difficult it becomes to take a collective view and to agree voluntarily to conserve the resource. There are no means by which individuals can in the future benefit from their own conservation activities since they cannot rely on others to do the same.
As a consequence, there is likely to be a scramble for those fish that are available. The dangers are probably greater for salmon than for most other species. Salmon are high value fish so the rewards to successful fishing are considerable. Equally important, salmon are relatively easy fish to catch in quantity because they are an

anadromous schooling fish. They can be taken easily in open waters and even more easily as they collect to enter their native estuaries and rivers.

Even when stocks are being artificially enhanced, if fishing effort is free to increase, the rate of exploitation may still be excessive. This is noticeable in British Columbia where the effects of a very successful and large public stock enhancement programme (see Chapter 4) have been offset by continuing excessive rates of exploitation in the waters through which the fish pass.

This tendency to overfish the common resource in unregulated public fisheries is often accompanied by problems of over capacity in the fishing fleet. In the early stages of the development of a fishery when maximum sustainable levels of catches are not yet achieved, the individual fisherman expects that the harder he works the greater will be his returns. If however he is seen to be successful, new boats will be attracted into the fishery. When this happens the resource becomes more heavily fished, returns per unit of effort will fall and the profits of each boat will be reduced. But each boat will still continue to fish as long as the rewards are greater than can be gained by working elsewhere.

In remote communities the opportunities for alternative work may be limited with the result that fishermen from these areas are prepared to remain in the fishery for lower returns than would be acceptable elsewhere, consequently increasing the pressure on the resource. Profits in good years also lead fishermen to expand capacity by buying more boats and gear. Once purchased any resale value is below the original cost and thus much of this investment can be regarded as a sunk cost so that the fishermen are prepared to remain in the fishery even if they are not earning enough to recover full capital costs. This tendency is exacerbated when fishermen are subsidised by loans and grants. Tax structures can cause similar outcomes. In bumper years in order to avoid heavy taxation, fishermen may instead choose to spend more heavily than would otherwise be the case on new equipment. As a result the industry may be over-capitalised, i.e. a higher level of capital equipment is employed than necessary to achieve optimum economic benefits from the fishery, and because capacity is excessive this

increases pressure on stocks as each boat tries to
increase its share of the catch. Without government regulation this can lead to the eventual
failure of the fishery.

3.3 HISTORIC RIGHTS AND SECTIONAL INTERESTS

Governments may have further objectives which
affect the strategies which they choose. Firstly, the government may wish to protect employment
or incomes in particular areas or for particular
groups of people, normally in areas where alternative employment opportunities are few and communities are very dependent on income from
fisheries. The maintenance of fishing rights for
these people may be seen as the best method of
ensuring this. It may be accepted that the costs
of these actions will be borne by other citizens.
  Governments may also recognise the historic
rights of particular groups to fish for salmon.
In some cases such rights are embodied in legislation of long standing: for instance in Scotland
some rights to fish for salmon have legal standing
over eight centuries old. In North America native
Indian communities have fought for and largely
succeeded in getting their rights to shares of the
salmon catch recognised by legislation.
  These rights involve international as well as
national issues. Disputes between the United
States and Canada in both Pacific and Atlantic
waters and between Greenland, the Faroes and other
Atlantic countries have centred around the rights
of individual countries to fish for salmon and
rights to a share of the total catch.

3.4 MANAGEMENT CONFLICTS AND THE MANAGEMENT PROBLEM

In the management of salmon fisheries·many governments have sought to achieve a combination of the
above objectives by attempting simultaneously to
conserve the resource, prevent over-capitalisation
and maintain jobs and rights to fish. This has
proved to be a difficult task and one of the
reasons for this lies with the conflicting implications for policy of the different objectives.
Attempts to reduce fishing effort can, at least in
the short term, have an adverse effect on incomes
in particular communities. Attempts to reduce

effort by reducing the numbers of fishermen have an adverse effect on employment levels, particularly serious in communities highly dependent on salmon fisheries. Attempts by one nation to conserve resources may lead to an unacceptable redistribution of catches to another nation less interested in such issues. The analysis of management policies below therefore shows that there have been many problems in managing the fisheries to achieve these multiple objectives. It should be noted that often these problems originate in the desire of governments to achieve multiple and conflicting objectives and not necessarily in the management tools themselves.

3.5 THE MANAGEMENT OF PUBLIC SALMON FISHERIES

Turning to management policies, the available management tools fall into three groups (see figure 3.1), each of which is considered in turn.

3.5.1   Controls on Entry

One of the oldest and most commonly used controls is to limit the right of free access to salmon fisheries by some form of restrictive licence or permit which has to be obtained before fishing. The objective is to reduce the number of individual units (boats or fishermen) allowed to fish and thus indirectly to reduce the catch taken. Ultimately, if the number of licences and the fishing capacity attached to them is sufficiently small in relation to the salmon stocks, such measures must be successful in reducing rates of exploitation to desired levels. Accordingly most fisheries do use some permits or licences for entry into fisheries.

In practice, however, no one has found it easy to use limited entry permits to solve the problem of excessive rates of exploitation. In most cases limited entry controls have been imposed after the problem of excessive rates of exploitation has already emerged. But to protect the interests of existing fishermen, typically they are all able to obtain licences which allow them to carry on fishing. The result is that capacity in the industry remains excessive even after limited entry is introduced. Governments have been reluctant, at least initially, to reduce the number of permits to a level consistent with

The Management of the Wild Resource

Figure 3.1: Management controls in public salmon fisheries

TYPE OF CONTROL	OBJECTIVES	EXAMPLES
a) CONTROLS OF ENTRY		
i) Licensing programmes to restrict numbers allowed to fish (limited entry)	restrict number of fishing units	most countries
ii) Buy-back programmes (purchase of boats and/or licences)	reduce number of fishing units	has in past been operated on small scale in British Columbia, Washington, and Oregon
iii) Licensing programmes with licence allocated for particular areas only	reduce numbers of fishing units, closer control of fishing effort, provide some direct incentive for conservation	Alaska, some European countries (under consideration in British Columbia)
b) CONTROLS ON GEAR		
i) Bans on fish traps	conserve stocks by reducing fishing efficiency protect fishermen against owners of fixed stations	most countries
ii) Bans on some gear and navigational aids eg. net mesh size, devices for finding fish limitations on vessel length and design	conserve stocks by reducing fishing efficiency	British Columbia, Washington, Oregon, Alaska, Greenland, Norway, Ireland, England, Canadian East Coast, Faroes

59

TYPE OF CONTROL	OBJECTIVES	EXAMPLES
c) CONTROLS ON EFFORT		
i) Area and time closures	conserve stocks, protect interests of other fishermen by limiting effort	most countries
ii) Quotas specifying size of permitted catch	conserve stocks by directly controlling landings	Greenland, Faroes, (under review British Columbia)
iii) Taxes on landed fish	reduce incentive to fish by reducing net rewards to fishermen. Proceeds may however be used to re-invest in the fishery eg through buy-back schemes	(under review British Columbia)
iv) Bans on salmon fishery in some areas	conserve stocks protect interests of fishermen in other areas	no open fishery for salmon in Scotland, some parts of England and Wales

adequate preservation of stocks because they do not wish to discriminate between individual fishermen in the fishery.

The best documented example of the problems that can occur is of the experiences of the British Columbia salmon fishery. The Canadian government has had a programme of licences to restrict entry into the fishery since the Davis Plan (3) of 1968. This was introduced with the dual objectives of increasing the earnings of fishermen and permitting more effective management of the salmon resource. The problems that occurred in implementing this limited entry programme were two-fold. Firstly, the initial numbers of licences issued were excessive. Secondly, like any other system of controls, once established, there was an immediate incentive for the fishermen to try to avoid them. Because there was a control on the number of boats but not on the amount of effort, people tried to avoid the controls by increasing the catching capacity of boats. Owners replaced existing boats with larger, more powerful and better equipped boats so that, although the number of boats did not increase, the catching capacity of the fleet did. In 1982, 14 years after the first limited entry controls, Commissioner Pearse in his investigation of the Canadian Pacific Fishery said:

Our catches of salmon (and roe herring) could be taken with fleets half their present size and half the cost now expended on fishing. If this were done, the value of the landings could well exceed the costs of harvesting in these fisheries by something in the order of $75-$100 million annually. (4)

These capacity increases had been encouraged by high and rising salmon prices in the 1970s but the increases in capacity did not disappear during the falling real prices of the 1980s. Thus:

In 1983 the salmon fleet suffered substantial losses of about $70 million on a total landed value of $107 million. The Pacific salmon fleet has an inordinately high debt load ... if the fishermen had made the required principal and interest payments on outstanding loans in 1983 they would have had only 10-15% of their total revenues left over to cover other costs. (4)

Pressures on the resource are still considered excessive (5). Similar problems exist in the salmon fisheries in Washington State, Oregon and California. Even in Alaska with its current high level of landings, in some sectors there are concerns about low returns to fishermen (6).

If it is decided, as has been the case in these fisheries, that the rights of existing fishermen are to be preserved and they are to be allocated licences, then a limited entry programme in itself cannot alone conserve the resource adequately. Supplementary controls will be needed and have in fact been introduced in most fisheries. In British Columbia as an example, there are regulations specifying the maximum size of boat permitted and the types of fishing gear which can be used. Such regulations are sensible in that they avoid excessive use of further capital in the fishery by preventing the fishermen from increasing the equipment employed. This is an efficient solution if the number of existing boats in the fishery is to be preserved but usually requires continuous monitoring of gear used. Note however as an alternative that it might well be possible for a much smaller number of boats using efficient gear to catch the same total number of fish at lower cost, as well as avoiding the heavy costs of policing the regulations. If a new fishery was being established this would probably be a more efficient solution. There are also problems in regulating the use of gear. Fishermen can find it difficult to fish selectively in terms of both quantity and species with the existing gear types available. But given the social and political constraints facing policy makers and the fact that the fishing fleet already exists, there may well be no realistic alternative.

3.5.2   Encouraging Exit: Buy-Back Programmes

Policies in British Columbia were not only based on licensing and gear restrictions. The Canadian Federal Authorities attempted to deal directly with the problem of the number of licences as well. They have implemented several relatively modest "buy-back" programmes where boats and their licences were bought out by the government. Similar programmes have been carried out in Washington and Oregon. These schemes are intended to reduce the number of boats in the

fishery by buying out the boat together with its
licence to fish.   It is the purchase of the
licence which is critical since it implies a
reduction in the number of people with rights to
fish.

None of the schemes have operated on a large
enough scale to have had a major effect on capacity and rates of exploitation of stocks.   Like
limited entry licences, however, if buy-back programmes are operated on a sufficient scale they
could reduce pressure on stocks by the desired
amount.   Moreover, they are equitable since they
compensate fishermen who leave the fishery.
Unfortunately it is the scale of the necessary
programmes which has so far inhibited their implementation since the sums involved are very
large.   In 1984 the Liberal Administration in
Canada proposed a reduction in capacity in the
Canadian Pacific fleet of 35-40% (1800 vessels)
which would have cost at least $100 million (5).
The economics involved were complex (7) and outcomes uncertain because it was not known what the
effect would be on fish prices, but the difficulties revolved around the question of who should
fund the programme.   It had been suggested (5)
that if it were to be funded by a tax on salmon
landings, then there would be no net cost to the
fishermen who would remain because they would
benefit directly from the reduced pressure on the
resource, i.e. the costs of the tax paid by the
fishermen would be offset by the increased value
of their landings.   It is also possible that some
of the tax burden could have been passed on in
higher costs to the consumer (although in competitive world markets this might have been difficult
to achieve).   But the larger and faster the
programme, the greater is the burden on present
fishermen and consumers in exchange for uncertain
future gains.   As a result fishermen may be
reluctant to co-operate.   An alternative is to
fund these programmes from general taxation.
Whether this is possible is largely a political
question and in the case of Canada where this
question has been most actively debated, a new
administration coming into power in 1984 did not
wish to follow through these earlier proposals, so
there is no actual experience of adjustments on
this scale.

Nevertheless, if funding questions can be
resolved, schemes like this do have a number of
attractions.   Firstly, by reducing capacity rather

than reducing the efficiency of operation of individual boats, these schemes should promote more rational use of capital equipment. Fishermen leaving the fishery are compensated for their loss of rights to fish and capital is released for more profitable purposes. If the remaining licences are transferable (i.e. they can be bought and sold) this should ensure that those expecting the greatest returns, in other words the most efficient, are the ones who do the fishing. Although it would reduce the numbers employed in the industry in the short-term, by maintaining the fishery and increasing its efficiency, more long term jobs will be retained than if through lack of action the fishery is allowed to fail.

3.5.3 Controls on Effort: Time and Area Closures

The controls mentioned so far act indirectly to control the actual effort and time spent on fishing but there are other more direct methods of controlling effort.

Most individual fisheries have controls on the areas and time periods in which salmon fishing is allowed. These controls have a long history. In Scotland, for instance, the earliest known legislation is an act attributed from 1179 (8) which made provision for a mid-stream gap in fishing weirs and provided for a weekly close time when all rivers were left free. Such time controls can take the form of statutory provisions as in Scotland or, as is the case in the United States and Canada, can give to government fishery agencies the powers to fix the details of opening and closing times during the season from the on-going results of stock monitoring. This can be a complex process. On the Pacific coast of North America, government agencies deal with regulations for several different types of fishery (seine, gillnet and troll) and for mixed stock fisheries where they must regulate simultaneously for different stocks in different states of abundance. They must also cope with the problems of controlling fisheries for the same salmon stocks but at different stages on their route back to their native rivers. Finally as well as making the controls effective, the administrators have to be seen to be fair to all the different groups involved.

The greater the fishing capacity relative to stocks the fiercer these restrictions have to be. In North American Pacific waters, some fisheries may be open for days or hours only and boats spend large proportions of their time idle even during the season - a considerable under-utilisation of capital resources. In the Atlantic the problems are so severe that some commercial sea fisheries are closed permanently. For instance in Scotland there is no high seas fishery for salmon and net fishing for salmon from boats is illegal. There is only one permitted legal inshore fishery on the East Coast of Canada.

The effect on fishing costs and on efficiency depends on whether the boats can be used in other fisheries when the salmon fishery is closed. Opportunities to do this vary, but where this is possible the effect of these closures on effi-ciency may not be serious. If no alternative opportunities exist, restriction may involve a more wasteful use of resources than if fewer boats fished for longer periods with consequent saving in administrative costs due to simpler regula-tions. However once more it comes back to objec-tives and to the fact that these restrictions are being imposed on an industry which already exists. If the boats and the licences are there anyway, such controls may within these constraints be the most efficient and often the only possible solution.

3.5.4   Controls on Effort: Quotas

To provide a solution which is simpler to admin-ister and control, quota systems operate in the Faroes salmon fishery. These have also been con-sidered for the Pacific fisheries of North America (9). Quota systems involve the setting of a total allowable catch (TAC) for a particular period of time. This total is divided up to give each licensed boat a maximum quota of fish which it is allowed to catch. This is supplementary to a limited entry licensing system and in theory it is the most direct and simple control of all. It removes the incentive for the individual boat to protect and increase its share of the catch. Instead it encourages the fishermen to adapt their vessels and fishing methods to take the permitted catch at the lowest cost. Controls on opening dates at the beginning of the season are still needed to prevent the taking of immature fish but there is no longer any need for complicated regulations on permitted gear and there is less need for detailed in season monitoring through openings and closures of the fishery.

Although such a system already works successfully in the Faroes fishery, it is unclear whether
such systems could work elsewhere. The problem
is simply that the accurate pre-season data on
which to base quotas does not always exist.
Quotas have to be set at the beginning of the
season but not enough is known about salmon eco-
systems to predict stock size accurately. There
is always a danger that quotas set turn out to be
excessive in relation to stocks in any one year
and in this case they would deplete rather than
conserve the resource. This problem has arisen
in Washington State fisheries where the native
Indian fishermen were given a quota based on 50%
of the estimated catch each year but where advance
estimates of catches were over-optimistic.
Conversely, too low a quota may reduce returns
unreasonably and will damage confidence in the
system. Furthermore, if quotas vary very widely
from season to season the incentive to reduce
catch costs may be reduced. The problem of quota
setting is particularly difficult for mixed stock
fisheries where different stocks may be in varying
conditions and this makes implementation of
schemes difficult in the North East Pacific where
many such fisheries exist. There is the further
problem of policing the quotas to ensure that they
are adhered to, which is in itself expensive.
Compulsory tagging, where the fishermen has to tag
each fish with his number before selling them,
might help, but these schemes have been criticised
on the grounds of their cost and are politically
very unpopular. Problems also exist in the
handling of salmon catches which are taken as by-
catches accidentally during the catching of other
fish.

Quotas do however have the advantage that they
regulate the size of the catch directly, which the
other controls mentioned do not. They can be
allocated between different types of fishing
according to a view as to what is equitable.
Therefore, given the problems which have been
experienced with other controls, there is
increasing interest in experimentation with quota
systems.

3.6 ALTERNATIVE MANAGEMENT SYSTEMS

The management problems in public fisheries have
their origin, as has been seen, in the common

resource problem which provides an incentive to deplete rather than conserve resources. Licences help to give fishermen an interest in conservation because they are in effect giving the fisherman private rights. However these are weak rights because there are so many licences that there is no incentive to take a long term view and to conserve the resource. As has been seen, other controls also pose difficulties. Yet another alternative is to try to create an incentive to conserve resources by giving the fisherman a more direct personal stake in the future of the fishery, rather similar to the private but far-sighted trap owner mentioned earlier.

One system of this sort is operated in Scotland. Here there is no legal off-shore net fishery so that the only fishing is land-based. Rights to fish from particular stations in estuaries and along the coastlines are held by private individuals or companies and can be bought, sold and leased fairly freely. By law and in theory the owners of the rights have joint responsibility with others in their district for the maintenance of the salmon stocks and the policing of the fishery in their area. The expenses of this again in theory are met by the individual rights holders in proportion to their stake in the fishery.

There are many problems in the Scottish salmon fishery. Like all the Atlantic salmon fisheries it faces the threat of resource depletion, not helped by the existence of illegal off-shore and river fishing which it is difficult to eradicate. There are also problems with the operation of the rights system mentioned above. The responsibilities of owners to look after the resource are more implicit than explicit in existing legislation. In some districts there are large numbers of rights holders and the sense of common interest and responsibility is weak. If survival rates and stock levels of fish are not known, nor are exact catches from different stations, then it is not clear what hatcheries and other facilities are needed and who should pay for them. There are also problems resulting from movement of returning fish along the coasts from estuary to estuary before finally entering their native river system. This complicates stock ownership issues. A consequence is that often in practice rights holders do very little. Further, one irresponsible owner can nullify the efforts of others and such owners do exist.

Nevertheless, it can be argued that this type of system could be developed to create a situation where owners of rights have a direct incentive to preserve the resource. If the numbers of holders of rights are small the individual rights holder can see a direct relationship between the effort to conserve and manage the resource and the marketable value of their fishing rights. Since the rights can be bought and sold, maintaining the value of the resource will increase the value of the rights. Obviously, there will never be a situation where all owners act responsibly, but the smaller the number of individual rights holders the greater the probability that they will identify their own best interests with those of the group as a whole. In Japan this type of system operates with fishing co-operatives who own their own hatcheries and have sole rights to harvest their returning fish in traps. This has been extremely successful, as indicated by the rapid increase in Japanese salmon landings (see Chapter 2).

It would be quite impossible to suggest land-based rights as a universal solution to salmon management problems. Firstly, international management issues (see below) preclude this in many cases. Secondly, Pacific salmon deteriorate as they return to freshwater and can be in poor condition by the time they reach the traps. A high percentage of the Japanese chum catch are lesser quality "dark" chums with a lower market value than fish caught out at sea. Troll fishing which catches fish further out produces the highest quality fish so this is a major advantage of this type of fishery. There is also the question of the fishing rights of troll boats which would disappear under a terminal fishing system, something that would certainly be unacceptable in North America.

Thus this is at best a partial solution and applicable only in some areas. However, it is possible to place more emphasis on terminal fisheries than is the case at present and to reduce some interception fishing, especially of mixed stocks. Alaska has moved in this direction by trying to identify individual licences with rights to fish in particular (albeit large) areas. Recent proposals in British Columbia have been somewhat similar. In Alaska there have been attempts to link these fisheries with hatchery programmes to give the fisherman a direct stake in and responsibility for enhancement within their area. The numbers of fishermen in each fishery are still large and the formation of the joint fishing/hatchery programmes is still too recent to comment on their long term potential. If, however, such programmes can lead the individual fisherman to identify more closely with the common interest in conserving the resource then the government regulatory problem is much eased.

3.7 INTERNATIONAL ISSUES

The problem of interception fishing of salmon stocks along the migratory route has already been mentioned. At this point it is appropriate to turn to the international dimension of salmon management since this is also bound up with the migration of salmon. The issue is about rights to interception fishing and rights to the resource. Some Pacific salmon returning to Canadian waters pass through Alaskan waters. Some salmon returning to United States rivers pass through Canadian waters. Canadian rivers discharge into Alaskan waters and so on. In the Atlantic, salmon from Northern Europe go to the Davis Strait off Greenland to feed, where they mingle with Canadian salmon which are also feeding in that area. Catches of these same stocks are subsequently made in Canada, the United States, Greenland, the Faroes, Iceland and all the countries of Northern Europe.

This poses a considerable management problem. Who is entitled to the resource? What can be done by one country if their resource is damaged by the failure of others to conserve it? For both Pacific and Atlantic salmon there have been problems in reaching international agreement on these issues.

The most serious problems at present are in the Atlantic salmon fishery. Atlantic salmon stocks are in a depressed state with falling catches over the past decade and international competition for the fish that are available has undoubtedly contributed to difficulties. The key to the international problem is the relationship between feeding grounds and spawning areas. On biological and economic grounds it is more

Figure 3.2:   Boundary river problems

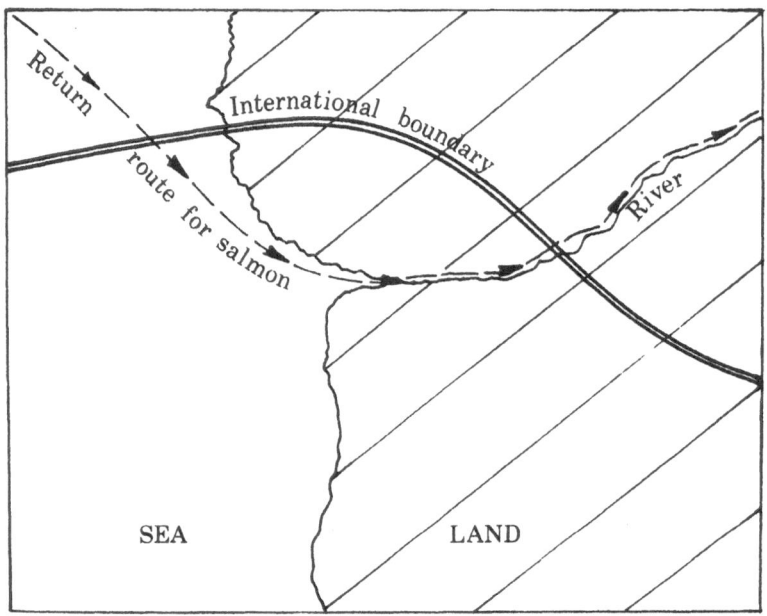

efficient to capture Atlantic salmon at the end of the migration. They are gaining weight up to this point and better stock management is possible since by this stage they have separated into separate river stocks whose individual management needs can be assessed. However, fish cross the Atlantic to feed and without these feeding grounds the stock could not be sustained anyway. The countries of origin have blamed those adjacent to feeding grounds (Greenland and the Faroes) for excessive fishing. Greenland and the Faroes have in turn blamed excessive drift net fishing and lack of environmental controls in the countries of origin and increasingly press a right to the resource at the point of greatest biological gain which is on the feeding grounds. Without the feeding grounds, they argue, there would be no fish returning to native rivers.

In the case of Atlantic salmon, the international solution has been a pragmatic one. Since each party needs the other to solve the common joint problem, then each must recognise the rights of the other. In 1982 a treaty was signed in Reykjavik establishing the North Atlantic Salmon Conservation Organisation (NASCO) (10). NASCO provides a forum for international information exchange, consultation and co-operation. It can commission research and make recommendations to its eight member governments, who include most nations in the North Atlantic with salmon interests. One of its first activities was to establish an agreed quota on Faroes landings to supplement the earlier quota established on landings from Greenland. In the founding treaty NASCO has undertaken to take into account in its recommendations not only: "the extent to which the salmon stocks concerned feed in the areas of fisheries jurisdiction of the respective parties", but also: "the interests of communities which are particularly dependent on salmon industries" (11). In other words, by guaranteeing the future of the Greenland and Faroes fishery on the feeding grounds, the Convention is buying the voluntary co-operation of Greenland and Faroes in conservation measures. In return Greenland and Faroes are expecting positive action from the countries of origin to control their drift net and illegal fisheries and to undertake other conservation measures. Controlling fishing by requiring dealers in salmon to obtain evidence as to the origin of the fish is also under discussion.

Figure 3.3: Migratory paths for Atlantic salmon

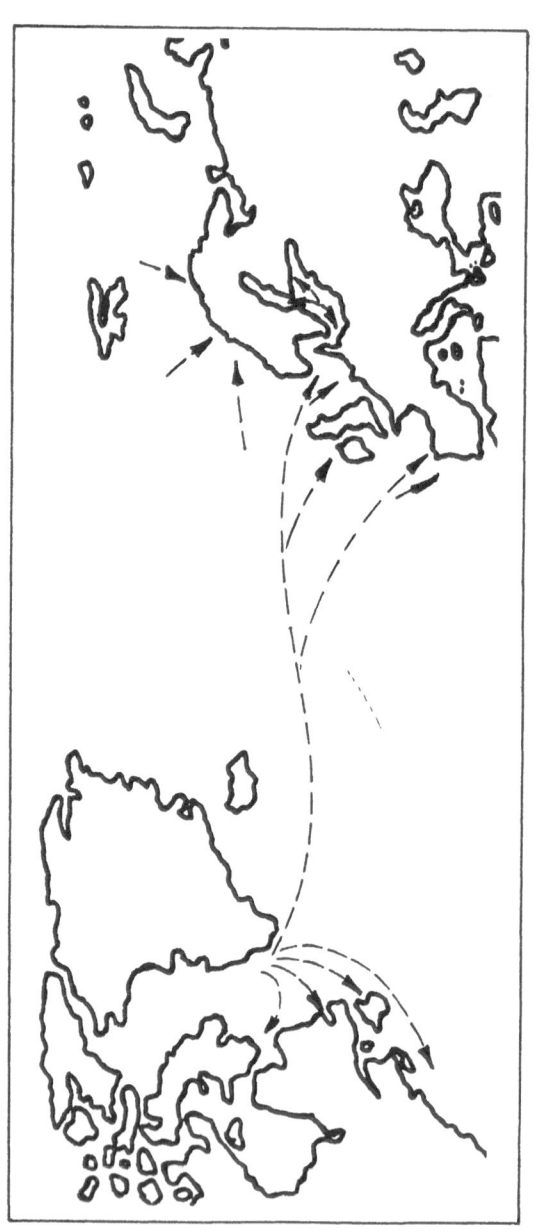

There are still difficulties. Available knowledge of stocks in the Atlantic is limited and
therefore stock assessment is difficult. Solving
disputes by using quotas can run into the problems
of excessive or inadequate rates of exploitation
discussed earlier. In spite of this, the new
body still represents a major step forward because
of the common international commitment to do something about the problems.

A similar treaty was agreed at the end of 1984
and ratified in 1985 between the United States and
Canada concerning the fishing of Pacific North
American waters. Until then although there was
international co-operation between the United
States and Canada in the management of the Fraser
River, agreement on other issues and areas had not
been reached. The new treaty has agreed an allo-
cation of the catch between the two nations in
order to ensure conservation of the resource. A
bilateral commission established by the treaty
will oversee intercepting fisheries in the United
States and Canada. The treaty has agreed on the
strategies to be used to rebuild chinook stocks
and has agreed proposals for the management of
transboundary rivers. Agreement between the
United States and Japan on the phasing out of
Japanese interception fishing in the Bering Sea
also appeared likely in mid-1986 (12).

Not surprisingly, international management
issues parallel the issues involved in the management of national fisheries. They involve trade-
offs of rights against the needs of specific
communities and the need to conserve the
resource. They require agreement on policy
methods and methods of stock assessment. What is
encouraging is that currently there appears to be
increasing awareness that the problems can only be
solved through international joint efforts and a
commitment to comply with common policies.
Nevertheless one of the ironies of this co-
operation is that by recognising feeding rights,
this co-operation is moving away from the
increased emphasis on terminal fisheries which is
occurring in some national policies. In practice
however such compromises are inevitable if the
interests of different parties are to be recon-
ciled. Once again solutions have to be seen to
be equitable before the common interest in conservation is served.

3.8 THE SPORT FISHERIES

While this book does not deal with the sport
fisheries for salmon, since the volumes taken in
sport fishing are large and the multiplier effect
on local economies considerable, the conservation
issue affects them as well.   This responsibility
has been recognised by many of the bodies responsible for sport fishing and by governments.
Pressures for stronger controls have come from
sports fishing interests, usually in the direction
of suggestions for greater emphasis on terminal
fisheries.   Because of the high value of the
sports fishery per fish caught, there is also more
willingness to meet the policing costs of
control.   In many areas the sports fishery is
more important than the commercial fisheries with
the result that the progress made on conservation
by sports fisheries is in fact very important in
the overall management of the resource.

3.9 OVERVIEW

It would be very misleading to claim in a brief
chapter to have done more than to introduce the
complex issues involved in salmon management and
the references given at the end of the chapter
cover the issues in more complete detail.   Some
general points can however be made in conclusion.   Firstly, and least controversially,
there is no alternative to controls of some sort
if salmon stocks are to be preserved.   Secondly,
if it is intended to preserve the livelihood of
particular groups of fishermen by giving large
numbers access to the resource, then there is no
simple method of regulation and high costs of
regulation are probably an inevitable consequence.   Simple solutions which drastically
restrict access to the fishery are often politically and socially unacceptable and on both
economic and social grounds the interests of
different types of fishing must be recognised,
with the consequent result of complicating management of fisheries.   Governments are increasingly
concerned about the high costs of regulation,
especially when borne by the general taxpayer, but
no easy alternative solution presents itself.
    As far as the future for salmon supplies is
concerned, this issue is discussed in more detail

in the next chapter. Given however the current general national and international recognition of the need for conservation and the lessons that have been learned from the workings of different types of policies, it is possible to be cautiously optimistic for the future.

NOTES

1. For more detailed treatment see T J Pitcher and P J B Hart, op.cit.
2. This is a simplified view of the monopoly case. For discussion see R Hannesson, Economics of Fisheries, Universitetsforlaget Bergen 1978. The assumption that monopolies will conserve the resource in fact depends much on the assumption they make about discount rates.
3. J Douglas MacDonald, The Public Regulation of Commercial Fisheries in Canada, Economic Council of Canada Technical Report No. 24, 1982, Ottawa.
4. The Commission on Pacific Fisheries Policy, P H Pearse, Commissioner, Turning the Tide: A New Policy for Canada's Pacific Fisheries, 1982, Vancouver.
5. Department of Fisheries and Oceans, Canada, A New Policy for Canada's Pacific Salmon Fisheries, 1984.
6. D M Larson, An Economic Profile of the Southeast Alaska Salmon Fishery, Final Report to the North Pacific Fishery Management Council, 1984.
7. D DeVoretz and R Schwindt, Harvesting Fish and Rents: A Partial Review of the Report of the Commission on Pacific Fisheries Policy.
8. R B Williamson, Inspector of Salmon and Freshwater Fisheries for Scotland, The Conservation of Salmon in Scotland, 1979.
9. A Scott and P A Neher (eds), The Public Regulation of Commercial Fisheries in Canada, Economic Council of Canada, 1981, Ottawa.
10. M. Windsor, A New Treaty to Manage Salmon in the North Atlantic, Greenwich Forum 11th Annual Conference, Edinburgh, 1985.
11. Convention for the Conservation of Salmon in the North Atlantic Ocean, Reykjavik, 1982.
12. Pacific Fishing, May 1986.

FURTHER READING

J A Butlin, <u>Economics and Resources Policy</u>,
    Longman 1981, London
Commercial Fisheries Entry Commission, State of
    Alaska, <u>Annual Report 1983</u>
Joint South East Alaska Regional Planning Teams,
    <u>Comprehensive Salmon Plan for South East
    Alaska</u>, Phase 1, 1981
J Karpoff, <u>Non-Pecuniary Benefits in Commercial
    Fishing: Empirical Findings from the Alaska
    Salmon Fisheries</u>, Commercial Fisheries Entry
    Commission, State of Alaska, 1983
D Mills, <u>Problems and Solutions in the Management
    of Open Seas Fisheries for Salmon</u>, Atlantic
    Salmon Trust 1983, Farnham
F Orth, J R Wilson, J A Richardson, S M Piddle,
    <u>Market Structure of the Alaska Seafood
    Industry, Volume 2: Finfish</u>, University of
    Alaska Sea Grant Report 78-14, 1981
A Scott and P A Neher (eds), <u>The Public Regulation
    of Commercial Fisheries in Canada</u>, Economic
    Council of Canada 1981, Ottawa
K Schelle and B Muse, Buyback of Fishing Rights in
    the U.S. and Canada: Implications for Alaska,
    <u>114th Annual Meeting of the American Fisheries
    Society</u>

Chapter Four

SALMON ENHANCEMENT AND RANCHING

4.1 INTRODUCTION

Salmon enhancement is the process by which wild salmon stocks, often under the possible threat of depletion, are supplemented by controlled hatchery production which is released into rivers, estuaries, or open sea waters, with the goal of supporting or building up these stocks. In most cases, the stocks used are of the same genetic source as the original stock in the system, they mingle with the wild stock, and are caught in the same salmon fishery as that for the wild stocks. Salmon ranching by comparison has been described by Thorpe as 'an aquaculture system in which juvenile fish are released to grow, unprotected, on natural foods in marine waters, from which they are harvested at marketable size'.(1)

To some extent, both operations are technically similar, and both derive from the techniques proposed and used by the naturalists and sport fishery owners of the nineteenth century, where young salmon stock were released into rivers in attempts to increase returns of adult fish. The main difference can be defined in the objectives: salmon enhancement is carried out with the aim of creating a public good, shared between many users, by national, international or regional agencies such as the Federal and State Government bodies of the USA, the Salmon Enhancement Programme of British Columbia and the Japanese fishery stocking programme (1)(2). Returns from enhancement programmes, although normally commercially realised, usually have social, environmental or political objectives in addition to commercial objectives.

Salmon ranching, by contrast, is normally

considered to be a more specific, targetted operation, usually with precise commercial objectives
benefiting a defined individual, private agency,
or collective party, in which a specific release
site and departure point is defined. The other
main difference is that salmon ranching, in its
ideal form, attempts to establish stocks separately from existing stocks, or more often, new
stocks where no stocks exist at present.

In practice, the technology and the factors
affecting the economics of production for the two
systems are essentially similar and will be
described as such. It is in the areas of economic return and economic management that the main
differences lie, and these will be discussed
separately.

The overall features of these operations are
described in Figure 4.1. As Table 4.1 shows,
both Atlantic and Pacific salmon species are used
and salmon ranching operations exist in North
America, Northern Europe, South America, the
Southern Oceans, and New Zealand. The important
difference between Atlantic and Pacific salmon is
that the latter can be released to the sea far
earlier in the life cycle than the former and as
will be shown, this can have important conse-
quences for the economic characteristics of the
operation (see also Chapter 5).

. The physical requirements for both operations
are the provision of a hatchery, normally on land,
a release point from which the young are introduced to the marine environment, normally in river
mouths, bays or estuarine coasts, and in the case
of ranching, a specific harvesting and restocking
point. In the simplest systems all of these
components are at the release point, and fish are
harvested as they enter release ponds or
ladders. A portion of the stock is selected for
breeding the next generation of young salmon, and
the remainder is harvested for sale. Figure 4.2
shows typical layouts.

Salmon enhancement operations have been best
developed and are best documented in the Pacific,
where Japan, USSR, USA, and Canada all contribute
stock to the salmon fishery. Salmon enhancement
is also carried out in the Atlantic, notably from
the Atlantic seaboard of Canada and the USA, from
Norway, Sweden and to some extent from the UK,
France, Spain and Portugal. The quantities of
released stock in the Pacific are immense; the
Fishery Rehabilitation, Enhancement and Develop-

Figure 4.1:   Production cycle in salmon ranching
              and enhancement

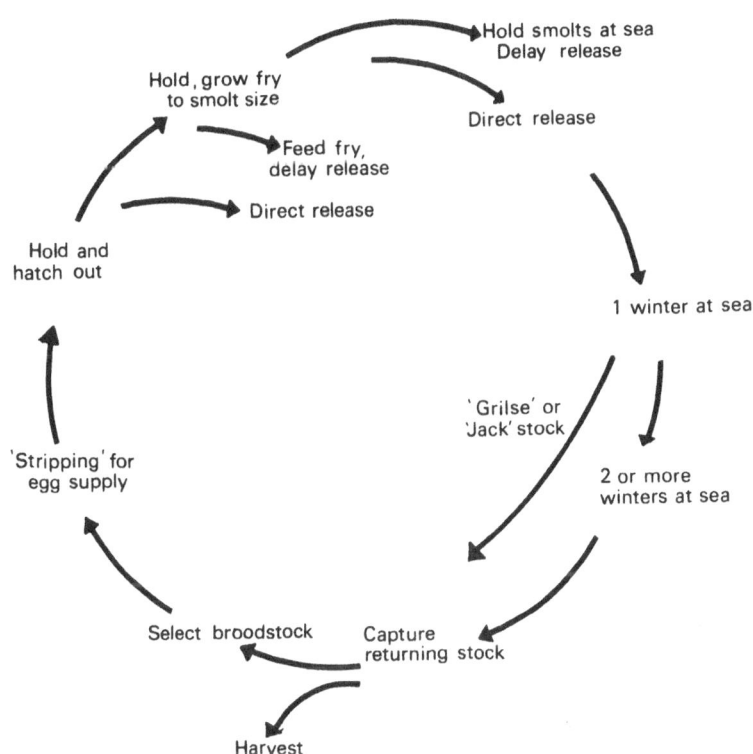

Table 4.1: Extent of salmon ranching and enhancement

Northern Hemisphere

USA	Pacific coast, Washington	All Pacific salmon,	Very large public enhancement programmes at state and federal level.
	Oregon, California	also S Gairdneri	Columbia River scheme pioneered techniques. Oregon main ranching area, large Weyerhauser unit (suspended) plus smaller private hatcheries.
	Alaska		Aquaculture associations predominate in ranching, normally with pink salmon. Extensive state programme with other species.
	Atlantic coast	Atlantic salmon	Limited state programmes of salmon release - also 'landlocked' stocks* of Atlantics, chinook, coho inland.
Canada	Pacific coast	All Pacific species, S. Gairdneri	SEP (Salmon enhancement programs) in BC, large and diverse, generally successful. No 'ranching' as such.
	Atlantic coast	Atlantic salmon	Federal and provincial programs of salmon enhancement. Recently cut back in funding level. Also 'landlocked' salmon stocks in Great Lakes.
Japan	Northern islands	Chum, cherry salmon	Large government hatchery programme, considered successful.
USSR	E Pacific	Pink, chum, coho	Large government hatchery programme.
	Arctic	Pink, chum	Experimental programme
Norway		Atlantic	Some river stock enhancement, compensation stocking.
Sweden		Atlantic	Extensive restocking of Baltic, compensation stocking of developed rivers
UK	Scottish	Atlantic	Some river stock enhancement, compensation stocking,
	Wales	S Trutta	plus small experimental 'ranching' programmes in W, NW Scotland.
Ireland		Atlantic	Large study of river stock enhancement programmes, also compensation stocking.
France	N W	Atlantic	River stock enhancement
Spain	N W	Atlantic	- -
Portugal	N	Atlantic	- -
Faroes		Atlantic	Experimental, semi-commercial salmon ranching
Iceland		Atlantic	Experimental, semi-commercial salmon ranching

Southern Hemisphere

Chile	Chinook, coho	Experimental, semi-commercial salmon ranching
Argentina	Chinook	Some experimental stock releases
New Zealand	Chinook (Quinnat)	Commercial salmon ranching
Tasmania	Atlantic, chinook	Experimental stock releases
Kerguelen Islands	Atlantic	Experimental stock releases

Figure 4.2: A generalised layout of capture systems

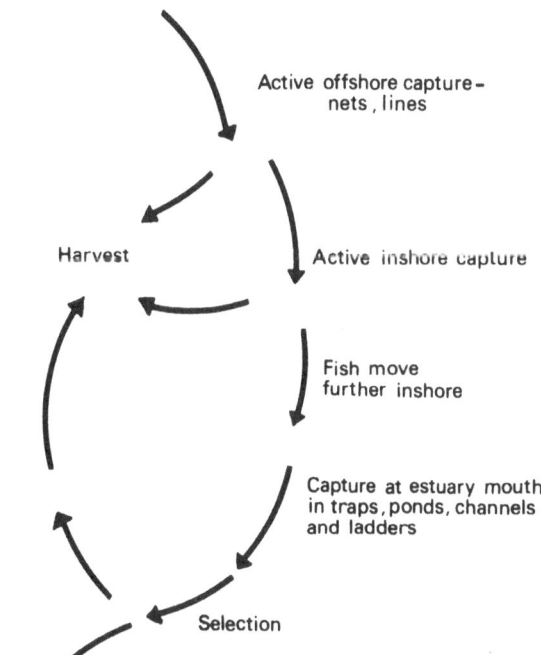

ment (FRED) in Alaska released almost 200 million young salmon in 1982, the ultimate aim being to supply more than 30% of Alaskan harvests, some 50 million fish, by this method. This would require a hatchery output of 340 million eggs.(3) The Salmonid Enhancement Programme (SEP) operating on a budget of $150 million Canadian for its first 5 year phase (1979-84) has a target of 50,000 tonnes of harvested fish.(2) The scale of operation in Washington, Oregon and California is also impressive, and exhaustive biological, technical and oceanic study has accompanied these develop ments. In Japan, the commercial chum salmon fishery is almost entirely produced from hatchery stock, as is much of the Soviet Union's pink and chum fishery.

Enhancement and ranching both have the signi- ficant advantage over salmon farming in that the major costs of feeding, holding and managing of stock from freshwater stage to harvest size are avoided because, at least theoretically, the oceans and the fish themselves provide the resources required. As will be shown, however, this has its price, and these theoretical advantages may not be so easily realised in practice.

4.2 CURRENT STATUS OF SALMON ENHANCEMENT

The main aims of the public enhancement programmes have been to build up salmon stocks to an acceptable level, normally corresponding to historic catch levels, modified or extended by estimates of the actual quantities of stock supportable by food resources in the ocean. In other localities, programmes have been initiated simply to restore a defunct fishery to any level.

Most of the large enhancement programmes, from their inception in the 1950s in association with hydroelectric developments such as the Columbia River Project in Washington State, have been relatively successful by most measurable terms. A considerable body of knowledge has been built up on the biology of the resource, and refined methods are now available to assess wild stock yields and hatchery reared stock yields, on the basis of better knowledge of the life cycle, the different river systems, of migration patterns and of catch effects.

In the North American programmes, many of the larger operations are now being trimmed by budget constraints; the lack of a specific target beneficiary has made accountability difficult to define, and the programmes difficult to defend politically. As a result, many of the programmes involve private groups more actively, new developments tend to be much more selective, and an increasing trend is seen towards private and/or group operated projects. Gordon (4) comments that on the face of increasing management complexity of traditional fisheries, there may be an evolution in the participation and management of the Pacific salmon resource, and that 'commercial hatcheries employing regional coastal based terminal fisheries or land based harvests may be a more biologically effective and cost-effective basis of resource management'.

Where analyses of costs (see following section) and benefits have been made, they have frequently been shown to be positive; the overall benefit to cost ratio of the SEP project (1982) was 1.3:1, with an average benefit:cost equivalent internal rate of return of 26%. It should be noted however that benefits quantified include commercial production, recreational fishery, and support of native food fishermen. As such, the methodology of defining benefits is open to other interpretation. Additional non-included benefits are jobs, direction of employment to less developed regions, and longer term protection of habitats. The FRED project in Alaska estimates a benefit to costs ratio of 2:1 to 3:1 in preliminary studies.

4.3 PRESENT DEVELOPMENT OF SALMON RANCHING

The development of salmon ranching is a relatively recent phenomenon and has been aided by two significant factors. The first is the technical factor of the generally greater level of confidence in early rearing technology and in the potential for the build up of salmon stocks in a specific location. This greater understanding of the operation of the system owes a great deal to the large enhancement programmes, particularly of Pacific salmon, and the significant quantity of biological data developed from them. As will be shown later, however, many operations are still extremely speculative since real biological (and economic) returns are little known.

The second major factor concerns the legal status of the operation, and the rights of ownership conferred on the stocks reared by the salmon rancher. Although significant changes have been made in a number of countries, this is perhaps the most difficult and often contentious aspect of salmon ranching - firstly because the reared fish are normally indistinguishable from wild fish, frequently forming part of the same general migration runs and thus often open to the same levels of access as the wild stock, and secondly because rights to catch salmon and controls over these rights (e.g. season, time, sizes) are often long-established, relate to the wild stock, and are frequently heritable property. There is however a sufficient number of cases either where legal rights are not too specifically defined (e.g. Southern Hemisphere areas with no tradition of salmonid resource management) or have been adapted sufficiently (e.g. New Zealand, Iceland), to permit some development, if only on an experimental scale.

In the cases where runs of salmon are being established where none previously existed, or where, say a single river system is owned and where the stock is reasonably separate from other regional stocks, the ownership issue can be resolved reasonably easily. In the more common case, however, where multiple or public ownerships of river systems are involved, where traditional fishing rights exist (e.g. commercial fishing licences, Indian territorial and fishing rights) and where stocks are being released from several different locations by different operators, the legal and political problems of establishing rights can be immense.

As will be discussed later and has already been mentioned in Chapter 3, the capture of returned fish at the release point provides perhaps the clearest biological indication of the origin of the stock and thus the least contro-versial means of harvesting, but it has important implications for stock quality, particularly with Pacific salmon, as the stock will no longer be 'ocean bright' and will not attract premium prices.

By far the most substantial development to date has been that in the Northern Pacific, where the easily-grown Pacific salmon species are being released in relatively large numbers in a range of private, co-operative and other operations. As Table 4.2 shows, release numbers and estimated returns are considerably larger than those else-where. The development of Quinnat (chinook) salmon ranching in New Zealand is perhaps the next most important activity, followed by the ranching of Atlantic salmon in Iceland and by the range of operations, mainly with Pacific salmon, in South America and other locations.

4.4 THE RELEASE AND RETURN MODEL

In the broad-spread public salmon enhancement operations a number of relatively sophisticated models have been built up with the aim of assess-ing biological returns to specific released stocks, of defining general and specific fishing effect on stocks, of relating overall stock survival and growth to external environmental factors, and of determining overall economic and social benefits to the programme costs (2, 5, 6).

At the level of the salmon-ranching operation

Table 4.2: Typical salmon ranching return rates

Species	Return rate	Notes
Atlantic	0.7%/3.9%	1 yr/2 yr smolts
	2.2-9.5%	2 yr smolts, Kollafjordur, Iceland.
(3%)	0.45/3.9%	Hatchery/Wild fish, Ireland, River Bush.
	2.3/8.3%	Mean 1971-1977, Ireland, hatchery/wild fish.
	4.2/7.9%	1 yr/2yr smolts, Mactaquac, Canada.
	2.5-3.0/10.0%	Reared/wild smolts, Norway.
	Some noted	Kerguelen Islands, salmon released into lakes and rivers.
Chinook	0.25-6.55%	16.2-64.9g size, Washington.
	1.8/1.5%	5g/11g fall Chinook, Washington.
(2%)	2.21±2.08%	Big Qualicum, B.C.
	3%	B.C. Hatcheries general.
	2.25-7.5%	Used by SEP Canada for planning.
Coho	10-15%	Saltwater release, Puget Sound.
	5.3-18.4%	Varying release methods, Puget Sound.
(2%)	8%/14%	Early/delayed release, Puget Sound.
	11.1 3.3%	Big Qualicum 1958-1971.
	15%	Used by SEP Canada for planning.
Chum	0.8-2.5%	Used by SEP Canada for planning.
(1%)		Commercial ranching returns, Oregon.
Pink	up to 16-17%	Prince William Sound, Alaska.
	2.7-4.6%	0.23-0.55g fry, Alaska.
(1%)	3.4%	Used by McNeil in return model.
	2.5-2.9%	Used by SEP Canada for planning.

Note: These reported returns are for indicative use only;
          distinctions between returns to specific sites and returns
          to a general fishery are not always made. There is as yet
          little long-term information in many of these
          circumstances.
Sources: SEP standards, personal communications, COST 46/4.
(   ):   Figure used in model.

and as the basis for defining effectiveness of enhancement techniques, the basic model is rather simpler, and concerns the relationships between the cost of growing and releasing stock and the value of the retained stock available for marketing. In the simplest sense this can be expressed in the following form:

Profit =   (Number of fish returning - number used as broodstock) x (sales value of fish), less the cost of growing and releasing the fry.

These figures can in turn be related to specific biologically defined characteristics, such as average weight, proportion of fish returning, and the proportion needed for broodstock. The relationship above can be developed further for the release of one or more different species.

$$P = \sum_{i}^{n} kN(1-m)\ WV - \sum_{i}^{n} LN$$

where   P = profit to operation e.g. dollars per year
           N = number of fish released
           k = biological return rate; ratio of number of fish returning/number of fish released
           L = hatchery cost function (non linear), e.g. costs per released fish
           m = proportion of returns used for broodstock
           V = average market value at point of capture, e.g. $/kg
           W = average weight of returned stock, e.g. kg
           n = number of different species produced.

   A model such as this suffers from a number of limitations - for example it assumes that W and k are independent of external factors, or of N itself, where it may well be the case that if a fixed feed resource is available offshore, increasing N beyond a certain point may reduce both W and k. Similarly it does not allow for competitive effects between the released species themselves. However, it does serve to illustrate the main functional relationship and given the uncertainty and imprecision attendant on identifying parameters in more complex models, this approach is often used as the basis for commercial planning.

A further factor entering commercial evaluation of release and return characteristics is the timing of returns; the time taken for the stock to return varies with species, location, and method of release. Generally, earlier returns produce lower average weights, but improve cash flow and allow the earlier selection of stock best suited for the local conditions. In the case of new stocks, the timing of return may also be critical in defining market opportunities, since returns during the peak catching periods of existing fisheries would normally be sold at the prevailing, lower, prices.

Table 4.3 summarises the typical characteristics of the main salmon species used for ranching, and defines biological return rates and average weights.

4.5 HATCHERY AND RELEASE OPERATIONS

Hatchery operations are generally similar to those of aquaculture production, save that the objectives - effective survival in the outside environment, good growth, and high propensity to return - are slightly different. At present, insufficient experience has been accumulated to control or influence the latter two of these, but the first appears to be effected by a combination of high-quality semi-natural growing conditions and of rigorous disease control. Thus the released fish should be given every opportunity for good early growth and conditioning with the minimum of physiological impairment through stress, disease or poor handling. Although the rearing techniques involved, requiring good facilities, clean water, and low stock densities may be more expensive than those in aquaculture, they are normally considered justifiable.

Table 4.4 describes the range of equipment and facilities used for rearing operations. The economic effect of this relatively conservative rearing policy depends much on the stage to which the fish are grown - in a simple fry release system such as that for pink and chum salmon (see figure 4.2), costs are comparable with those for

Table 4.3:   Characteristics of main species

Species	Distribution	Early Rearing	Release	Migration
Atlantic Salmon Salmo salar	N Atlantic, E Canada. Introduced to Puget Sound in cage culture	18(S1) or 30(S2) months Release in freshwater	release as smolt @ 20-25g+ (S1) or 35g+(S2) salmon	Offshore, oceanic feeding grounds. Return @ 12 months grilse or 24+ months
Sea Trout Salmo trutta	Widespread N America, S Hemisphere (also Falklands)	18,30,42 mths Release fry in freshwater	Release as smolt as above	Inshore waters return 12,24,36 months
Steelhead Salmo gairdneri	N America some Europe	18,30 months	Release as pre- smolt or smolt @ 20-25g	Offshore or midrange waters, 1-3 yrs.
Brook Trout Salvolinus fontinalia	N America	6-8 months	Fry into streams or small release	Short-sea migration
Arctic char Salvelinus alpinus	N America N Europe	6-8 months	Fry into streams or small release	Short-sea migration
Chinook Salmon Oncorhynchus tshawytscha	N Pacific New Zealand and some in Chile	Varying times	Normally release 18 month smolt 20g+. Occ. 6 mth 1.5g	Offshore, but can localise with delayed release, 1-5 yrs at sea.
Coho Salmon Oncorhynchus kisutch	N Pacific some in France Chile	18 or 30 months	Normally 18 mths smolts @ 20g+ Delayed release to 30g+	Offshore, but can localise with delayed release. Jacks return 12M, others 24 months.
Pink Salmon Oncorhynchus gorbuscha	N Pacific, also W USSR/North Norway	5-8 weeks	Fry, fed or unfed at 1g+	Offshore, return after 18 months
Chum Salmon Oncorhynchus keta	N Pacific, also Chile	5-8 weeks varying time downstream	Fry fed or unfed	Late release stay inshore, then off- shore 2-3yrs at sea.
Sockeye Salmon Oncorhynchus nerka	N Pacific	18 months, fry into lakes	Into lakes	12-30 months at sea offshore
Masu Salmon Oncorhynchus masou	N W Pacific	5-7 months	Smolt release or fry	1-2 years

Table 4.3 continued

Typical size	Quality	Market	Other
Grilse 1-4kg Salmon 3-5kg	Good	Upper levels well developed	Salmon return earlier in year than grilse. May spawn more than once.
0.3kg @ 12 mths 1.5kg @ 24 mths 3.0kg @ 36 mths	Good	Lower than Atlantic salmon Not developed	Sea-running brown trout. Some may return later same year as release.
1.5-6kg	Good	Not well developed Local catches. Sport fish.	Sea-running rainbow trout. May spawn again. Salmo mykiss related species in W Pacific.
0.5-3kg	Med- good	Mainly sport fish	
1.0-3kg	Med- good	Mainly sport fish	Well adjusted for cold climates.
4kg+ up to 15 kg+	Good	Upper levels	Die on spawning. Silver colour/red flesh most prized. Also known as spring, king.
Jacks 1kg+ Minimum 3kg+	Medium	Middle levels	Die on spawning. Also known as silver salmon.
2 kg+	Low- medium	Lower levels	Die on spawning. Fixed 2-year cycle. Also known as humpback.
3kg+	Low- medium	Lower levels	Die on spawning. Also known as keta, dog salmon.
3kg+	Medium	Low-middle levels	Die on spawning. Also known as red salmon.
1.0kg-2.0kg		Mainly Japanese market	Also known as cherry salmon.

Table 4.4: Hatchery facilities

Type	Normal Species	Loading	Other
Heath tray	All	3-9,000/tray	Use netting, media for fry at loading, otherwise hatching only.
Atkins box	Pink, Chum	300,000/box	2 cells per box
Pallant box	Pink, Chum	150,000/box	Gravel media/screens
Quinsam box	Pink, Chum	750,000/box	Maximum. gravel media, now use plastic media
Zinger box	Pink Chum	150,000/tray	2-4 trays/stock, Alaskan use
Plastic trays	All	3-12,000/tray	Stacked in boxes or raceways
GRP troughs	Atlantic	10,000/trough	
GRP tanks	Atlantic	10-30,000	First feeding
Kepper channels	Chum	600,000/channel	Covered, gravel media
Capilano troughs	Chinook, coho	55,000/trough	First feeding

aquaculture production, while producing Atlantic
salmon smolts to the highest quality is likely to
result in higher costs. However, while this
'high quality' approach is generally followed,
there is as yet little conclusive proof that a
simpler, cheaper production approach would result
in substantially reduced return performance. At
the simplest level, fry-rearing hatcheries can be
operated with the minimum of attention; stock is
placed in the hatchery units, the eggs hatch, the
fry emerge when ready and move downstream to seawater. In the more conventional and widely

Figure 4.3: Utilisation of hatchery facilities

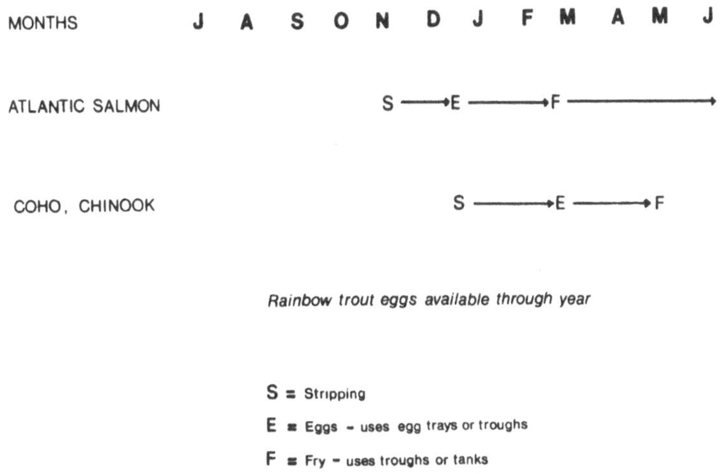

MONTHS	J	A	S	O	N	D	J	F	M	A	M	J

ATLANTIC SALMON   S ——→E ————→F ————————————→

COHO, CHINOOK   S ————→E ————→F

Rainbow trout eggs available through year

S = Stripping

E = Eggs – uses egg trays or troughs

F = Fry – uses troughs or tanks

used hatcheries, because fry production is highly seasonal, there is considerable underutilisation of facilities and labour, whereas in smolt production the hatchery operates year-round. The useful season can be extended by combining several species or by combining fry release with smolt release. This also improves utilisation of facilities, as Figure 4.3 demonstrates.

Although very small release units have been developed, it is usually considered that a certain minimum release number is desirable, both to conserve a viable biological stock at sea and simply to ensure a reasonable number of potential broodstock returning to build up subsequent generations. Table 4.5 shows typical minimum release quantities, although it should be noted that these are guidelines only.

The release operation offers the operator the greatest area of control over the return performance of the stock. Release points can vary from hatching boxes from which fry emerge into streams, to release ponds and ladders, to cages or enclosures in estuary or marine sites. The size at which the fish are released and the timing of release relative to the normal stock life cycle

Table 4.5:   Minimum release numbers for
              establishing return runs

Species	Minimum Number	Notes
Pink salmon	200,000 to 500,000	Alaskan practice
	100,000	Small size operations in BC (British Columbia)
Chum salmon	200,000+	Alaska
	100,000	Small operations in BC
Atlantic salmon	10,000	Sweden
	1,000+	Iceland
	5,000	Faroes
Chinook salmon	5-25,000	Small operations in BC
	20,000	Alaskan practice
Coho salmon	15-20,000	Small operations in BC
	20,000	Alaskan practice

Sources: SEP standards, personal communications, COST 46/4

Note: These are not scientifically defined minimum, rather the smallest numbers released in practice.

can both have a significant effect on the extent of outward migration and the timing, weight and rate of return.   Thus, in general, smaller fish, released relatively early, tend to migrate further into the open ocean paths and to return later, normally at a higher average weight, but to have a lower return rate to the point of release.   Fish held for longer and perhaps fed for several weeks or months in release ponds or cages tend to travel within relatively short distances.

Indeed it may be possible to develop stocks within well defined areas – for example larger

Table 4.6: Choices faced by the operator

Early release

Advantages	Disadvantages
Cheaper, simpler production method	Higher losses, lower numerical returns
No feed supplies required	Apparently less easy to target
Easier to release in several locations	stock to specific release area.
Possible larger returns phase, larger size.	

Late release

Better stock 'targeting' may be possible	Higher cost of rearing, more complex facilities
Better numerical returns in appropriate conditions.	Feeding required
	May need marine site
	Possible increased numbers of small, early-maturing fish

sounds or bays where fish grow reasonably quickly and return early, at a lower average weight but with a higher return rate. However, for both coho and Atlantic salmon this approach often favours the return of 'jacks' or grilse, respectively. In this case the market price is also significantly lower, while the cost of holding stock for late release is significantly increased.
Table 4.6 illustrates the choices faced by the intending operator. In many locations, however, the effects of these alternatives are not yet fully defined - indeed different strategies may yield entirely different results in different circumstances.

4.6 RETURNS AND HARVESTS

As Figure 4.4 illustrates, there are a number of factors, most of which are beyond the control of the operator, which may influence final return rates. Unlike salmon farming, there is little control over the pattern of return, the size ranges, or the quality of the fish. While table 4.7 shows the kind of return pattern which might be expected, it is quite possible for distribution to be more concentrated, with say 80 % of

Figure 4.4:   Factors affecting final returns

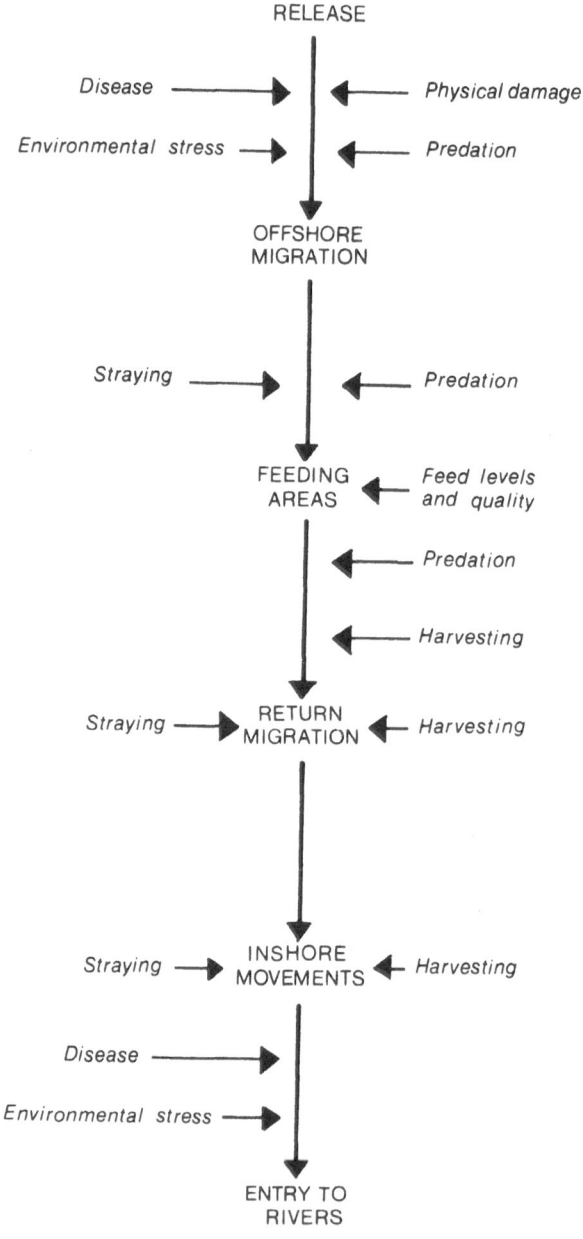

Table 4.7a: Species return and build-up model

Years post-release	1	2	3	4	5	TOTAL
Atlantic						
No %	35.0	55.0	10.0	–	–	100%
Wt kg	2.5	3.5	4.5	–	–	av 3.25 kg
Wt	87.5	192.5	45.0	–	–	325
% wt	26.9	59.2	13.9	–	–	100%
Chinook						
No %	–	15.0	65.0	25.0	5.0	100%
Wt kg	–	4.5	9.0	12.5	15.0	av 10.40 kg
Wt	–	675.0	585.0	312.5	75.0	1040
% wt	–	6.75	56.3	30.0	7.2	100%
Chinook Delayed Release						
No %	5.0	5.5	55.5	35.5	–	100%
Wt kg	2.5	4.5	9.0	12.5	–	6.38 kg
Wt	12.5	247.5	315.0	62.5	–	637.5
% wt	2.0	38.8	49.4	9.8	–	100%
Coho Normal Release						
No %	10.0	90.0	–	–	–	100%
Wt kg	2.5	3.8	–	–	–	3.67 kg
Wt	25.0	342.0	–	–	–	367
% wt	6.8	93.2	–	–	–	100%
Coho Delayed Release						
No %	20.0	40.0	40.0	–	–	100%
Wt kg	0.5	2.5	4.0	–	–	2.7 kg
Wt	10.0	100.0	150.0	–	–	270
% wt	3.7	37.0	59.3	–	–	100%
Pinks						
No %	–	100.0	–	–	–	100%
Wt kg	–	2.0	–	–	–	2.0 kg
Wt	–	200.0	–	–	–	200
% wt	–	100.0	–	–	–	100%
Chum						
No %	–	15.0	65.0	20.0	–	100%
Wt kg	–	2.5	3.5	4.5	–	3.55 kg
Wt	–	37.5	227.5	90.0	–	355
% wt	–	10.6	64.0	25.4	–	100%

Table 4.7b: Salmon production cycles

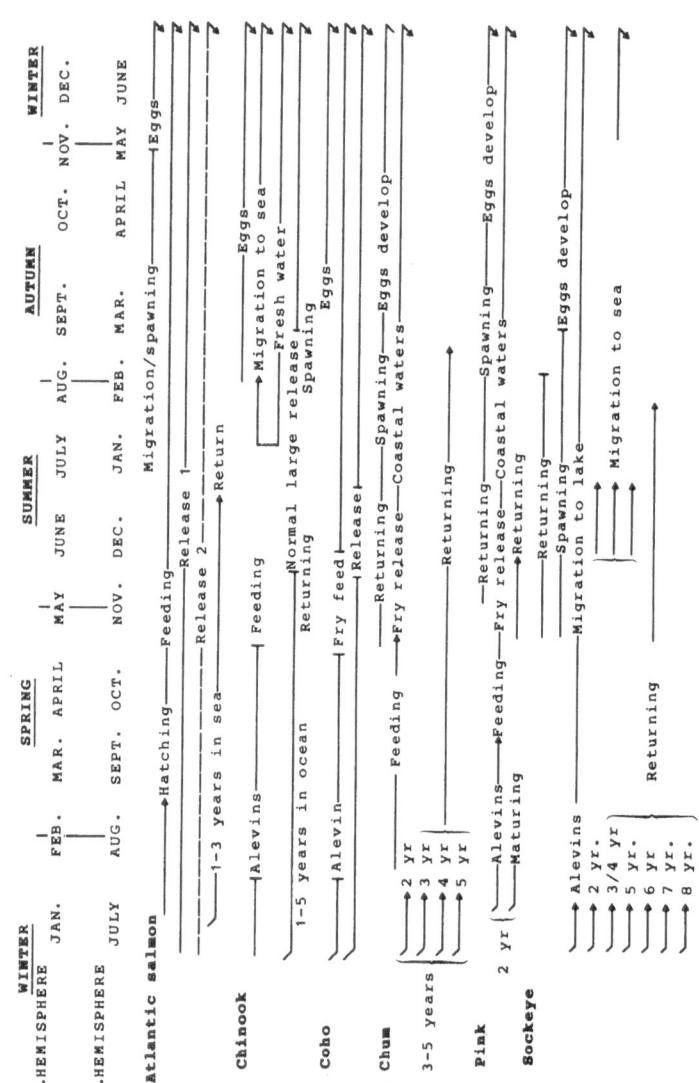

the returns in a salmon ranching system coming within one week. The operation is thus more closely akin to fishing, and normally takes place in the same annual period. In the case of enhancement, returns are expected to be at the same time as that of the native wild stock. Therefore, unlike salmon farming, enhancement and ranching do not have the same potential for planning supply and distribution in line with market needs, particularly for fresh products.

The question of returns becomes particularly complex where the fish form part of a mixed stock, which is fished at several points before their final coastal destination. In the case of private, commercial salmon ranchers, rights to fish are often confined to the release area only, rather than to the open sea fishery. Even in these circumstances, commercial fishermen often protest about the possible risk to other stocks returning with the ranched stock. Thus in Oregon, pressure has been applied to introduce compulsory tagging – usually a microtag injected into the nasal area of the fish. This and its decoding equipment are expensive, but have been found not to impair performance and can be used to identify ranched fish (McNeill, personal communication). The rancher, however, rarely has a right to tagged fish caught in the open sea.

In the case of the "Aquaculture Associations" such as those in Alaska, by definition private, non-profit organisations and normally collectives of fishermen, processors, or other interested groups, the issue is relatively straightforward. In these cases groups normally operate the fishery in the area, and returns effectively support or enhance the fishery with sales of any surplus over broodstock requirements taken at the release/ recapture point and used to offset hatchery running costs. The latter are however financed principally by a levy on landings.

The actual return and harvest facilities are generally quite simple, ranging from 'active' systems such as sweep, ring, or gill netting, to 'passive' systems such as set nets, ladders, or holding pools.

In the case of systems near or on the mouths of rivers, where the fish (particularly Pacific salmon) are nearing spawning condition, a system of holding pools and broodstock selection is operated at the same time as the harvest. A labour-intensive activity, the broodstock are stripped of eggs and milt to form the basis of the next generation's production.

4.7 CAPITAL AND OPERATING COST CHARACTERISTICS

It is even harder to generalise about the financial characteristics of salmon enhancement and ranching operations, than in the case of salmon farming, in that the variables are not generally controllable, the sites and types of operation are diverse, and the period of development, particularly in ranching has been relatively short. However, the main capital and operating costs of typical operations can be defined to some extent, by comparison with equivalent aquaculture operations. Agencies such as SEP have also detailed analyses of release costs from specific facilities.

Historically, certain operations have been developed in less than ideal circumstances, and both capital and operating costs are higher than expected by current aquaculture standards. Likewise, the public hatchery operations have tended to be 'over-engineered' by commercial standards and produce significantly higher production costs.

Table 4.8:   Typical levels of cost[1]

	Atlantic smolt		Pacific fry		
Numbers of fish	200,000	500,000	1,000,000	5,000,000	20,000,000
Cost per 1000 fish[2] direct release	£ 630	£ 510	£ 140	£ 80	£ 60
Production costs per 1000 fish delayed release	1000[3]	900[3]	150[4]	90[4]	70[4]

1 See also chapter 5, table 5.10.
2 Costs are highly dependent on site characteristics.
3 Costs of seawater held smolts.
4 Typical costs of fed fry.

Tables 4.8 and 4.9 show approximate levels of cost expected in enhancement and ranching operations and demonstrate the effects of species choice, system and size. However, although economies of scale can be demonstrated for particular sets of site conditions, changes in site development costs, release and return system design or harvesting operations, can easily over-shadow these effects.

Table 4.9: Effects of species choice

	Stock cost per 1000	Return rate	Average size	Theoretical production costs per kg[1]
Atlantic salmon	£500	3%	3.0 kg	£6.25
Chinook	£100	2%	3.5 kg	£1.89
Coho	£100	2%	3.0 kg	£2.15
Chum	£ 90	1.5%	2.5 kg	£3.07
Pink	£ 90	3%	2.0 kg	£1.91

1   Assume £0.30/kg harvesting cost, with 0.2% broodstock requirements.

Applying the simple production model described earlier in the chapter, a target area of minimum biological return rates can be defined (Table 4.10). By comparison with Table 4.2 earlier, it can be seen that many species and systems are potentially capable of yielding a favourable economic return and there is therefore a good incentive for development.

An examination of the sensitivity of a typical operation demonstrates that biological return rates substantially .control profitability, and that changes in individual input costs have relatively little effect (Table 4.11). In addition to implying a reasonable scope for higher hatchery costs in areas of good biological return, this

Table 4.10: Minimum biological return rates required

Species (average return weight kg)	1.0	1.5	2.0	2.5	3.0	3.5	4.0
% return rates required							
Species Costs[1]	%	%	%	%	%	%	%
Atlantic							
£800	22.8	15.2	11.4	9.1	7.6	6.5	5.7
£600	17.1	11.4	8.6	6.8	5.7	4.9	4.3
Chinook							
£150	4.3	2.9	2.2	1.7	1.4	1.2	1.1
£ 60	1.7	1.1	0.9	0.7	0.6	0.5	0.4
Coho							
£120	6.0	4.0	3.0	2.4	2.0	1.7	1.5
£ 60	3.0	2.0	1.5	1.2	1.0	0.9	0.8
Chum							
£120	12.0	8.0	6.0	4.8	4.0	3.5	3.0
£ 60	6.0	4.0	3.0	2.4	2.0	1.8	1.5
Pink							
£ 80	8.0	5.3	4.0	3.2	2.7	2.4	2.0
£ 60	6.0	4.0	3.0	2.4	2.0	1.8	1.5
Sockeye							
£100	10.0	6.4	5.0	4.0	3.2	3.0	2.5
£ 60	6.0	4.0	3.0	2.4	2.0	1.8	1.5

1   Costs per 1000 fish released.   Cost alternatives depend on
production route chosen, see table 4.8.

Assuming values of marketed fish per tonne: Atlantic £3500,
Chinook £3500, Coho £2000, Chum £2000, Pink £1000, Sockeye
£1000.

Table 4.11: Sensitivity of profit to return rate

£ gross profit per kilo for varying return rates

Return Rates %	0.5	1	2	5	10	20
£ gross profit per kilo						
Atlantic	(63.1)	(19.0)	(7.6)	(0.7)	1.5	2.5
Chinook	(10.8)	(1.9)	1.1	2.6	3.1	3.3
Coho	(14.0)	(4.0)	(0.7)	1.0	1.5	1.8
Chum	(9.0)	(4.5)	(0.7)	0.4	0.7	0.8
Pink	(9.0)	(4.5)	(0.7)	0.4	0.7	0.8
Sockeye	(5.7)	(1.5)	(0.1)	0.6	0.8	0.9

(   )   losses

Assume 0.2% use for broodstock.

Based on following values and costs:

	Market price per tonne £	Size returning fish kg.	Cost per 1000 released fish £
Atlantic	3500	3.0	600
Chinook	3500	3.5	150
Coho	2000	2.5	120
Chum	1000	2.0	60
Pink	1000	2.0	60
Sockeye	1000	3.0	60

also implies that technical improvements leading to lowered hatchery costs are unlikely to have highly significant effects.

4.8 THE BUILD-UP TO PRODUCTION

One of the most important financial features of a salmon ranching operation, shown also to some extent with salmon-farming (see Chapter 5), is the long time required for financial returns to build up to a commercially profitable level. There are two factors in this, firstly the time required for the harvestable fish to complete their seaward cycle and return to the release point and secondly the longer-term effect that it may take several generations of fish to establish a population capable of returning at sufficient levels to support production costs. Thus it is only by self-selection, breeding from returning fish, that the overall 'focusing' of stock to the release site may be established.

The effect of this on overall stock returns is illustrated in Table 4.12. When this is combined with the typical capital and operating cost profiles described in the previous section, the overall cash-flow characteristics of the operation can be defined (Table 4.13). Table 4.13 shows the effect of different sets of assumptions concerning return rates and production value. Thus it can be seen that although minimum biological return rates may be within the levels observed with particular stocks, the rates required to cover both the risks and the initial cash-flow requirements are rather higher.

In the case of systems currently in operation, it is interesting to examine the expected production costs of the harvested fish. As Table 4.14 shows, the application of a 15% biological return rate in pink salmon production - reported in certain Alaskan operations, results in a production cost of £0.78 per kg. A 10% return of Atlantic salmon results in a production cost of £2.61 per kg.

4.9 CONSTRAINTS

Perhaps the main constraint at present is the need to prove the commercial viability of salmon ranching, and the need for clearer definition of

Table 4.12: Build-up to production - 1000kg final output

	Years post release														
	0	1	2	3	4	5	6	7	8	9	10	11	12	13	14
Atlantic															
2.5kg	-	27	54	81	108	135	161	167	191	215	219	269	269		
3.5kg	-	-	59	118	178	237	296	355	414	474	533	592	592		
4.5kg	-	-	-	14	28	42	56	70	83	97	111	125	139		
TOTAL	-	27	113	213	313	413	513	613	713	813	913	986	1000		
Chinook - normal															
4.5kg	-	-	7	13	20	26	33	39	46	52	59	65	65	65	65
9.0kg	-	-	-	56	113	169	225	282	338	394	450	507	563	563	563
12.5kg	-	-	-	-	30	60	90	120	150	180	210	240	270	300	330
15.0kg	-	-	-	-	-	7	14	22	29	36	43	50	58	65	72
TOTAL	-	-	7	69	162	262	362	462	562	662	762	862	956	993	1000
Chinook - delayed release															
2.5kg	-	2	4	6	8	10	12	14	16	18	20	20	20	20	
4.5kg	-	-	39	78	116	155	194	233	272	310	349	388	388	388	
9.0kg	-	-	-	49	99	148	198	247	296	346	395	455	494	494	
12.5kg	-	-	-	-	10	20	29	39	49	59	69	78	88	98	
TOTAL	-	2	4	133	233	333	433	533	633	733	833	931	990	1000	
Coho - normal															
2.5kg	-	7	14	20	27	34	41	48	54	61	68	68			
3.8kg	-	-	93	186	280	373	466	559	652	746	839	932			
TOTAL	-	7	107	207	307	407	507	607	707	807	907	1000			
Coho - delayed release															
0.5kg	4	7	11	15	19	22	26	30	33	37	37	37			
2.5kg	-	37	74	111	148	185	222	259	296	333	370	370			
4.0kg	-	-	59	119	178	237	297	356	415	474	534	593			
TOTAL	4	44	144	244	344	444	544	644	744	844	941	1000			
Pink															
2.0kg	-	-	100	200	300	400	500	600	700	800	900	1000			
Chum															
2.5kg	-	-	11	21	32	42	53	64	74	85	95	106	106		
3.5kg	-	-	-	64	128	192	256	320	384	448	512	576	640		
4.5kg	-	-	-	-	25	51	76	102	127	152	178	203	229		
TOTAL	-	-	11	85	185	285	385	485	585	685	785	885	975		

Table 4.13:   Cash flow characteristics: salmon ranching

£ sterling

	0	1	2	3	4	5	6	7	8	9	10	11	12	13
Atlantic salmon[1]														
System costs	40.0	40.0	25.0	-	-	-	-	-	10.0	10.0	-	-	-	-
Variable costs	5.0	13.0	13.0	14.0	15.0	15.0	15.0	15.0	15.0	15.0	15.0	15.0	15.0	15.00
Fixed/SVC*	15.0	33.0	36.0	36.0	36.0	36.0	36.0	36.0	36.0	36.0	36.0	36.0	36.0	36.00
Total costs	60.0	86.0	74.0	50.0	51.0	51.0	51.0	51.0	61.0	61.0	51.0	51.0	51.0	51.00
Sales revenue	-	-	-	2.1	9.5	17.9	26.3	34.7	43.1	51.5	59.9	68.3	76.7	82.6
Net revenue	(60.0)	(86.0)	(74.0)	(47.9)	(41.5)	(33.1)	(24.7)	(16.3)	(17.9)	(9.5)	8.9	17.3	25.7	31.6
Pacific salmon[2]														
System costs	50.0	50.0	14.0	-	-	-	-	-	10.0	10.0	-	-	-	-
Variable costs	5.0	15.0	29.0	29.0	29.0	29.0	29.0	29.0	29.0	29.0	29.0	29.0	29.0	29.0
Fixed/SVC*	25.0	30.0	69.5	69.5	69.5	69.5	69.5	69.5	69.5	69.5	69.5	69.5	69.5	69.5
Total costs	80.0	95.0	112.6	98.5	98.5	98.5	98.5	98.5	108.5	108.5	98.5	98.5	98.5	98.5
Sales revenue	-	-	-	-	250.0	500.0	750.0	750.0	750.0	750.0	750.0	750.0	750.0	750.0
Net revenue	(80.0)	(95.0)	(112.6)	(98.5)	151.5	401.5	651.5	651.5	641.5	641.5	651.5	651.5	651.5	651.5

1   At 8% final return level, 3.0kg average weight; release 100,000 smolts/year. Returns based on build-up shown in table 4.12. This represents a 'poor to average' case. Sales price £3500 per tonne.

2   At 10% final return, 3.0kg average weight; release 1,000,000 fry/year, with returns building up more rapidly than 4.12; 33% year 4, 66% year 5, 100% year 6 onwards. This represents an 'average to good' case. Sales price £2500 per tonne; based on coho/sockeye.

*   Semi-variable costs

Table 4.14.: Expected production costs: current ranching operations

	Atlantic salmon	Pacific salmon
Biological return rate	10%	15%
Hatchery size	200,000	1,000,000
Return size	3.0kg	2.5kg
Spawners	0.5%	0.5%
Return weight	57.0t	362.5t
Total hatchery operating cost	126,170	139,400
Production cost per kg*	£2.61	£0.78

*Allowing for 0.40/kg for catching, icing, bulk packing, etc.

the site locations and operating conditions
required to achieve sufficiently high biological
return rates. Considering the suggested minimum
release size earlier, even the smallest operation
may represent a sizeable development cost, with no
significant guarantee of longer-term financial
returns.

It is too early to be able to define the overall performance of this section of the industry,
as many of the earlier relatively poor results may
be attributed to deficiencies now known, but it is
significant that a number of salmon ranching
operations have disposed of their facilities,
including one of the largest and most ambitious
schemes, that of the Weyerhauser Corporation near
Springfield, Oregon.

For salmon ranching to develop further in the
future, therefore, requires the prospect of a substantial return on investment to offset the
apparent risk. Should some of the more recently
developed operations prove their long-term viability, current perceptions could change significantly.

It is likely, however, that as production
requirements become more closely defined, site
availability will decrease correspondingly.
While at present it may be possible to identify
many locations with good, easily developed, freshwater supply, access to a good release area, with
good offshore feeding grounds and limited interception, these could rapidly be used up as the
industry develops.

Additionally, where several release operations
feed into the same offshore grounds, the areas of
conflict referred to earlier become intensified,
and the potential for specific biological
targeting of stock may diminish. In particular,
while a separate and active fishing industry
continues, there will continue to be political and
legal problems in supporting a separate private
salmon ranching industry.

Finally, the short harvest season of the
ranching system imposes a significant marketing
constraint, and may effectively deprive operators
of the means to influence or develop markets.

4.10 LONGER TERM DEVELOPMENTS

Perhaps the most significant developments are
currently occurring in the Southern Hemisphere on

the Chilean coast of South America and on the South Island of New Zealand. A combination of excellent environmental conditions and the absence of a significant competing commercial fishery results in an excellent potential for viable production.

As these and most other operations are still in the early development phase, it is difficult to define the final return levels, though results appear to be encouraging enough to stimulate further, expansion. Given suitable biological return rates, a salmon-ranching system is relatively insensitive to individual release cost variables, and is likely to compete effectively in relatively adverse economic conditions.

In the longer term, the political and legal questions concerning 'grazing rights' are likely to become more sharply defined, particularly as life cycle pattern, stock characteristics, and environmental effects become more clearly identified. It is entirely conceivable that with the exception of those areas where little competing activity occurs, private salmon ranching operators may become absorbed within fishery enhancement systems of the Aquaculture Association form, which may in turn take an increasing rôle in overall salmon production, replacing some of the public enhancement operations.

REFERENCES

1. J E Thorpe (ed), <u>Salmon Ranching</u>. 1980, Academic Press. London and New York

2. Department of Fisheries and Oceans, Government of Canada/Province of British Columbia, <u>Salmonid Enhancement Program Annual Report 1981</u>, Vancouver

3. Alaska Department of Fish and Game, Division of Fisheries Rehabilitation, Enhancement and Development, <u>1982 Annual Report to the Alaskan State Legislature</u>, FRED Report Series No. 2, Juneau

4. M R Gordon, <u>Private Sector Involvement in Pacific Salmon Enhancement</u>, Industry Information Report No. 7, November 1982. B.C. Research Fisheries Technology Division, Vancouver

5. R M Peterman, Model of salmon cage culture and its use in preseason forecasting and studies of marine survival. <u>Canadian Journal of Fisheries and Aquatic Science</u>, 39, 1982

6. J McDonald, The stock concept and its application to British Columbia salmon fisheries, <u>Canadian Journal of Fisheries and Aquatic Science</u>, 38, 1981

FURTHER READING

W J McNeil and J E Bailey, <u>Salmon Rancher's</u>
    <u>Manual</u>, 1975.   Northwest Fisheries Centre,
    Auka Bay Fisheries Laboratory, Processed
    Report, National Marine Fisheries Service,
    NOSA, Alaska
COST 46/4 workshop, <u>Sea Ranching of Atlantic</u>
    <u>Salmon</u>,   Commission   of   the   European
    Communities 1982
M K Farwell and T R Porter, Atlantic Salmon
    Enhancement Techniques in Newfoundland, FAO
    Technical Conference on Aquaculture, 1976.

Chapter Five

SALMON AQUACULTURE

5.1 INTRODUCTION

The aquaculture of salmon holds with it the potential of changing output independently of the
natural and other constraints of the wild salmon
resource, of controlling the output quality and
timing, and of defining a different pattern of
cost structure for the product. It is the purpose of this chapter to demonstrate the way in
which aquaculture alters the production characteristics for salmon, to define the constraints
present and to present a description of the
economics of production.

5.2 SALMON AQUACULTURE PRODUCTION

By far the greatest current production is of
Atlantic salmon, though limited quantities of
Pacific salmon, mainly chinook and coho, are grown
in British Columbia, Washington State, in New
Zealand and in Brittany (Table 5.1). Production
of Atlantic and Pacific salmon in the British
Columbia area is expected to increase significantly in the next few years, and other areas are
also likely to commence production. There is
also a limited amount of landlocked salmon production, mainly of Atlantic salmon.
     Of the aquaculture producers, Norway is
clearly dominant, followed by Scotland, although
other producers are rapidly developing. Most of
the technological features of modern Atlantic
salmon aquaculture are however based on these two
producers, and it is on the production characteristics of these that the chapter will be based.
The production of salmon on the Pacific coast is
discussed separately because it exhibits some
differences.

Table 5.1: Production of farmed salmon

Country	Production, tonnes				Notes
	1975	1980	1985	1990	
Norway			22,196	50,000	Site availability constraints, also environment/disease constraints
Sweden	–	–	Neg.	500	Mainly trout. Some salmon for restocking
Finland	–	–	Neg.	200	Mainly trout, some salmon
Scotland	Neg	1,000	6,900	20,000	
Ireland	–	500	650	3,000	
France	–	Neg.	Neg.	500	Limited production coho salmon
Iceland	–	Neg.	200	3,000	
Faroes	–	Neg.	1,000	15,000	
Canada	50	100	500		Excellent site potential. Mainly coho and chinook, some Atlantic. Mainly West Coast
USA	50	100	400		Constraints due to environmental controls.
Chile	–	Neg.			Excellent site potential. Coho and chinook
New Zealand	–	Neg.			Excellent site potential, mainly chinook. Ranching also.
Australia	–	Neg.	Neg.		Tasmania – chinook. Also ranching.

neg. = negligible

5.3 METHODS OF PRODUCTION

Technically, the aquaculture system attempts to replicate or adapt the natural lifecycle and habitat characteristics of the salmon, but with management control being exercised over as much as possible of the production cycle. Further technical descriptions may be found in Edwards (1), Sutterlin and Maxwell (2), Mowinckel and Kvalheim (3). The process may be summarised as follows (see also Figure 5.1):

a) Egg production   This uses good quality, well oxygenated fresh water with an ideal temperature of approximately 8°C, within a range of 2-12°C. Broodstock fish are 'stripped' of eggs and milt; the fertilised eggs are normally developed in troughs or in baskets, trays or boxes. The eggs are sold or transferred at either 'green', i.e. newly fertilised, level, or 'eyed', i.e. when larval eye is present. Hatcheries range from small part-time operated units producing 1000-5000 eggs per annum, to large suppliers of half a million to 10 million eggs.

b) Fry and smolt production   Hatched eggs are grown in tanks, troughs, raceways and increasingly in net cages in freshwater, to the stage at which fish are ready to adapt to seawater (6-18 months with Pacific species, 18 months or 30 months with Atlantic). The ideal temperature is 10 to 12°C and production is highly seasonal, the fish usually being ready for seawater by late spring. However some producers modify temperature and/or photoperiod to accelerate adaptation (see Thorpe (4)). Feed is normally of dry pellet type, graded upwards in size as fish develop. Feeds are normally fishmeal based, and provide a feed to fish weight conversion of about 1.6-2.1. With Atlantic salmon, 18 months (S1) smolts, ready for sea, are about 20-50 grammes in weight and 30 month (S2) smolts are about 50-200 grammes. Production units range from about 10,000 smolts per year to 1 million or more.

c) Ongrowing   Smolts are introduced either into cages, enclosures, or onshore tank systems, and then grown in seawater to

115

Figure 5.1: Salmon farming production cycle

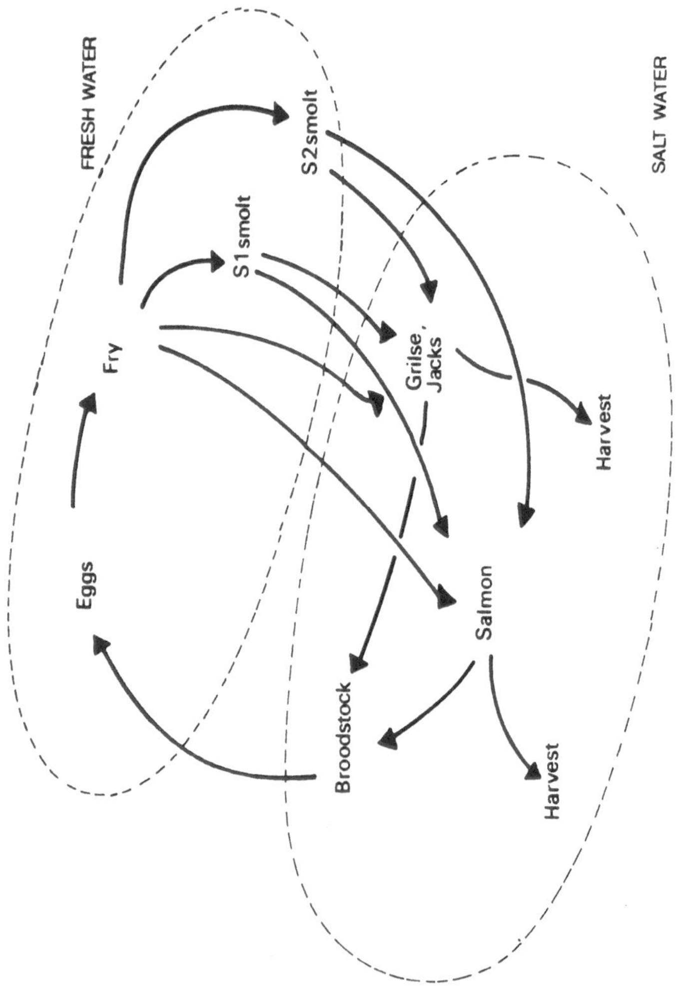

market size. This can be either as pan-
sized fish (e.g. coho), grilse (Atlantic
salmon marketed after one winter at sea –
normally 1-3 kg), or as salmon (sometimes
called two-winter fish – normally 2-5
kg). Feeds are predominantly dry pellets,
though a substantial quantity of freshly
produced moist diet is also used. Food
conversion averages are about 1.8-2:1 for
dry foods and 2.5-4:1 for moist foods.
Ideal water temperatures are 12-14°C
although there is some suggestion that cold
winter temperatures stimulate overall
growth. Individual holding units range
from 100 cubic metres to 5000 cubic metres
or more and stocking densities range from
10 to 40 kg per cubic metre of water.
Production units range in size from units
producing 30 tonnes to 1000 tonnes or more
per annum.

d) Broodstock These are taken either from
maturing farmed fish or from wild caught
supplies. Broodstock have normally spent
two or three winters at sea and are
stripped either at sea or onshore in fresh
water, to provide the new stock.

Many of these procedures, particularly the
first two described (egg production and fry and
smolt production), are common to salmon aquaculture, to stock enhancement and to salmon
ranching (see Chapter 4). In all of these,
production methods are essentially similar, and
production units may serve more than one purpose. In salmon farming there are both specialist producers concentrating on eggs, smolts or
ongrowing, and integrated operations controlling
the entire life cycle. There is also often substantial trade between specialist and integrated
sectors. This diversity is frequently the result
of the different site requirements for the
different life cycle stages, though many integrated operations use several sites, often well
separated geographically.

5.4 MANAGEMENT OF PRODUCTION PROCESSES

5.4.1 Key parameters
Although the fundamental techniques of salmonid
culture have been developed over at least 100

117

years, aquaculture production is still largely
based on the natural seasonal characteristics of
the fish and thus the extent of manipulation of
fish size or time of production is still at best
limited.

The extent and significance of management can
be summarised in the sections 5.4.2-5.4.6 below.

5.4.2   Control of intake

There is a certain natural variation in popula-
tions of salmonids which can be used to some
extent to widen the time period over which eggs
are taken, and fry and smolt are produced.
However the intake to the sea is still normally
within an 8 to 10 week period from May to July,
and there may be no direct effect on timing of
final market size production.   The most useful
effect is in improving the ability to supply a
range of sizes during as much as possible of the
production year.   One particular problem however
is that large, early stocked smolts, although
growing more rapidly, frequently mature early as
grilse, thus achieving a reduced marketable weight.

5.4.3   Control of growth

Temperature control is used by some producers,
particularly those in higher latitudes, to
increase the proportion of S1 smolts, thus reducing the overall holding time in the hatchery.   In
freshwater a limited number of producers attempt
to accelerate production of smolts ahead of the
normal period.   Thus in North Norway, one farm
releases fish routinely at 20 grammes, usually
within 8 months of hatching.   Opportunities for
using temperature control (e.g. by heating and
re-using water) for ongrowing to market size are
limited in view of water requirements and the
consequent cost (see e.g. Muir (5)); as a result
of this the possibilities then become limited to
that of site selection for the required temperature regime.   As already mentioned however,
accelerated growth in seawater often causes the
risk of early maturation.

5.4.4   Control of holding conditions

There is some evidence that continuous water flow
such as that produced in intensive flowing water
tank systems may stimulate growth, and thus
improve productivity.   Some circumstantial evidence also suggests that the use of larger holding
facilities, whether tanks or cages, may also

confer growth advantages. While in the first case this has yet to be clearly established in production conditions, as other factors may mask the effects of flow, many cage-based producers are now moving to the use of larger facilities, in the expectation of improving growth rates in the larger volumes provided.

5.4.5 Feeding

The efficacy of feed input is a major management factor, both in maintaining and stimulating growth, and in controlling costs. As shown later, feed costs contribute significantly to variable operating costs. Effective management aims to maintain the lowest FCR (food conversion ratio) consistent with growth. Considerable changes in feed quality have come about in the last 10 years; quality control in particular has improved to the point where feed performance can be better standardised, and where 'growth programming' in which stock weights are planned ahead by controlling feed input, may become routinely feasible.

5.4.6 Harvest decisions

There are three main harvest sizes in salmon aquaculture.

Pan-sized production This is only used for coho salmon, where the relatively low freshwater rearing costs permit the marketing of fish in the 200-500g range, usually after 6 months in sea. In cooler waters these are grown over the summer months, with an autumn harvest, while in warmer waters such as the Brittany coast, late stock are used for winter rearing and spring harvest (6).

Grilse Grilse are defined as Atlantic salmon maturing after 1 winter at sea, in a size range of 1-3kg. If the stock is left too late in the summer, · flesh quality deteriorates, colour changes and there is greater competition from wild supplies, resulting in very poor prices. Thus it is normally decided to grade out and sell as early as possible, typically between March and June, even though some growth may be foregone. In warmer water operations, where high growth rates are achievable and a large proportion of stock mature early (e.g. Ireland and France) the objective is usually to aim

for large grilse sizes, and concentrate on the
use of a 12 month cycle of seawater
production. In colder waters, grilse ratios
are typically 10-30% and grilse are normally
graded out, the remaining stocks being grown
over two winters to provide the bulk of the
larger fish.

In cases where very low grilse ratios occur
(e.g. less than 5% with certain stocks), a
decision may be taken not to harvest the
grilse, but to leave them undisturbed to
mature, then to reabsorb their gonads (eggs or
milt) and recondition over the following
winter. However, as these stocks are
particularly prone to disease, the risk of
loss and the possible infection of other
stocks may outweigh any advantages of reduced
handling earlier in the production cycle.

Salmon As Atlantic salmon smolts are
currently (1986) priced at around £1.50 each,
the cost of initial stock is highly
significant in the overall production cost of
smaller sizes of fish (see Figure 5.2). It

Figure 5.2: The significance of smolt cost in
             salmon production cost

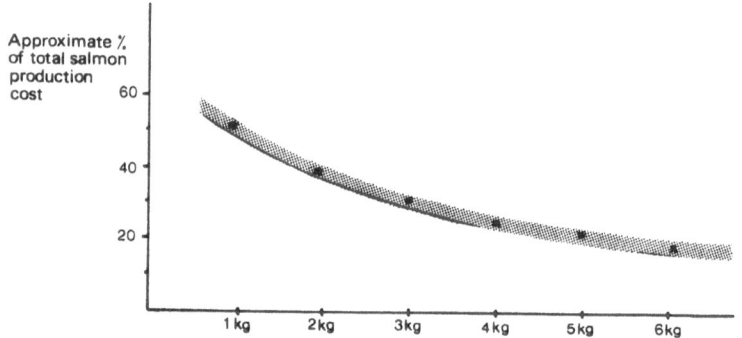

is thus a predominant objective to maximise growth and average size. As already noted, this is achieved in most cases by holding stock into and through a further winter. Salmon stocks are normally sold from October/ November through to March/April, depending on market status and stock performance.

Most recently however, changing patterns of market demand are reducing the potential for production of larger fish. Thus at least in UK markets, optimum sizes appear to be those corresponding to large grilse or small salmon (1.5-3.5kg). Figure 5.3 illustrates the typical 'production year' for salmon farmers.

Figure 5.3: The salmon-family "production year"

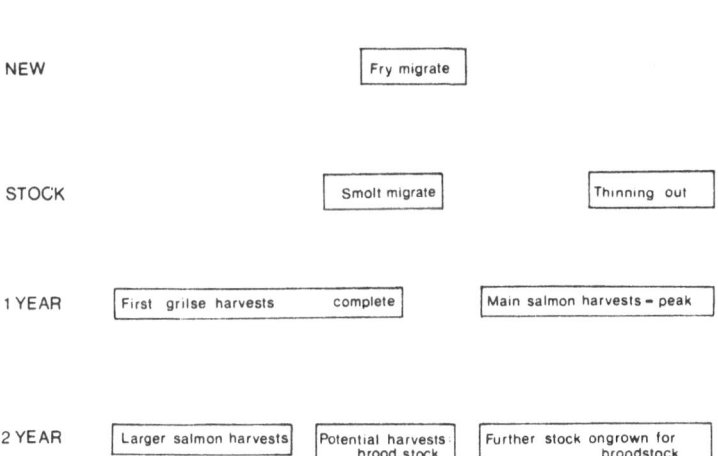

5.5 CAPITAL COSTS OF PRODUCTION

5.5.1 Scope
The cases of both freshwater and seawater stages will be considered, the latter described for both sea cage and pumped seawater systems; reference

is also made to enclosure systems. Specific
capital costs may vary considerably according to
site characteristics, and while it is normally the
objective to select sites offering the lowest
development costs, other reasons such as access to
markets, good water circulation, availability of
land, or development incentives, may dictate a
choice of other sites with higher than normal
costs.

5.5.2 Freshwater production

Table 5.2 illustrates the main features of the
capital costs of freshwater production. For
Atlantic salmon after the initial egg-rearing
stage, there are two main options for production:
freshwater cages, or tank systems. Costs are
described for both, at levels of 20,000 and
200,000 smolts annual output; with mixed S1, S2
stocks.

Table 5.3 shows the capital cost effects of
major production alternatives. Thus it can be
seen that cage systems, where sites are available,
are generally the cheapest form of production,
while heated and recycled systems are generally
more expensive. The economies of scale which
occur are generally related to the costs of water
supply development and staff accommodation, as
these become relatively less important as unit
size increases. In practice, many sites are
limited by water resource availability; for
Atlantic salmon, development beyond 200,000 smolts
is usually achieved by the use of multiple sites,
each with associated infrastructure costs.

In these circumstances recycle systems, by
increasing efficiency of water usage, may become
more competitive as management and infrastructure
costs can be concentrated on a single, larger,
site which is not constrained to the same degree
by the availability of water.

In the case of Pacific salmon, the complexity
and cost of production depends on whether the
species is grown to fry or smolt stage. As all
of the Pacific species can be introduced to the
sea at fry stage, most commercial operators do so,
thereby saving substantially on capital cost.
Table 5.4 shows outline capital costs for
production levels of 50,000, 200,000, 1,000,000
and 5,000,000 fry. As water resource

Table 5.2: Main features of freshwater production of smolts: capital costs for 20,000, 100,000 and 200,000 smolts, mixed S1, S2

| | Annual production of smolts | | | |
	20,000	100,000	100,000[1]	200,000
	£	£	£	£
Site preparation	1,500	3,500	3,500	6,000
Water supply	1,500	10,000	10,000	15,000
Hatchery building	2,500	10,000	10,000	20,000
Accommodation	-[2]	35,000	35,000	35,000
Egg trays	600	3,000	2,700	6,000
Fry tanks	2,600	13,000	11,700	26,000
Smolt tanks	2,400	12,000	14,400	24,000
Plumbing installation	2,000	8,000	9,000	16,000
Vehicles, etc.	-[2]	6,000	6,000	7,000
Services	200[2]	3,000	3,000	5,000
Alarm/Monitor	200	1,000	1,000	1,500
Miscellaneous	500	2,000	2,000	3,000
Contingencies (10%)	1,500	10,650	10,830	16,400
Total	16,500	117,150	119,130	180,400
Ratio	0.825	1.172	0.902	1.191

1 Using higher S2 ratio - 50% S2, as opposed to 30% S2 in base case.

2 Assuming minimal requirements - supplied by operator.

Table 5.3:   Effects of the production alternatives
on   freshwater   capital   cost: 100,000
smolts

£ Sterling

	Base case	Freshwater cages	Recycle System
Site preparation	3,500	1,000	3,500
Water supply	10,000	1,500	2,500
Hatchery building	10,000	10,000	10,000
Accommodation	35,000	35,000	35,000
Egg trays	3,000	3,000	3,000
Fry tanks	13,000	3,000	13,000
Smolt tanks	12,000	-	12,000
Installation	8,000	2,000	12,000
Cages, moorings, nets	-	20,000	-
Boat/service gear		5,000	
Vehicles, etc.	6,000	6,000	6,000
Services	3,000	1,000	4,000
Alarm/Monitor	1,000	200	1,500
Water treatment	-	-	10,000
Backup	-	-	4,000
Miscellaneous	2,000	2,000	3,000
Contingencies	10,650	8,770	11,950
Total	117,150	96,470	131,450

Table 5.4: Outline capital costs: Pacific salmon fry[3]

| | Annual production of fry | | | |
	50,000	100,000	1,000,000	5,000,000
	£	£	£	£
Site preparation	300	500	4,000	15,000
Water supply	200[1]	1,000	10,000	15,000
Hatchery building	200	1,000	10,000	40,000
Accommodation	-[2]	10,000	35,000	50,000
Egg trays	500	1,000	10,000	40,000
Fry tanks	500	2,000	20,000	90,000
Installation	500	1,000	8,000	35,000
Vehicles, etc.	-[2]	-[2]	6,000	12,000
Services	-[2]	-[2]	10,000	15,000
Alarm/monitor	-	-	2,000	3,000
Water treatment	-	-	1,000	2,000
Backup	-	-	5,000	10,000
Miscellaneous	100	200	10,000	20,000
Contingencies (10%)	230	1,670	13,100	34,700
Total	2,530	18,370	144,100	381,700

1　Simple on-stream incubation base

2　Operator provides.

3　Ready for sea.

requirements are correspondingly lower, so the limitations of water supply become less evident, and individual units producing large quantities of fry become possible, without the use of water recycling.

Beyond the basic infrastructure costs such as staff housing, power supplies, access and water, economies of scale are not likely to be significant. Greater economies can occur if mixed species production is used, allowing several different stocks to be grown in the same facilities.

5.5.3   Seawater production
This is normally the most significant sector in both capital and operating costs, as by far the greatest weight increase and stock capacity occurs in the seawater phase.   For Atlantic salmon, a weight ratio of about 60:1 can be assumed (2.4kg harvest weight: 40 gramme smolt weight), for Pacific salmon, about 300:1 (2.4kg harvest, 8g fry).

The most common, and the cheapest form of seawater production for both Atlantic and Pacific salmon is cage culture.   This is illustrated in Table 5.5 which shows typical cage sizes currently in use, and gives appropriate unit costs.   Table 5.6 describes typical capital cost profiles for production levels of 50t and 200t.   These figures are applicable generally to both Pacific and Atlantic salmon, although, as will be discussed later, the influence of growth rates and harvesting schedules and hence the ratio of annual production to installed capacity, can have a significant influence on costs.

Generally, as with freshwater production, once initial infrastructure provision is made, no substantial economies of scale have been observed.   As in freshwater, higher production levels are normally accommodated by multiplying the numbers of sites.   As producers become increasingly aware of the risks of environmental contamination, the use of multiple sites has increased.

On the smallest scale, it is notable that by assuming the availability of existing accommodation, existing storage, vehicle and boat, not financed by the farm, which is typical of family-operated project, capital costs are significantly reduced.

The increasing pressure of demand for sites of

Table 5.5: Typical salmon cage dimensions

Type	Dimension (l x b x d)[1] metres	Approx. volume cubic metres
Standard Scottish Timber cage	7 x 7 x 4.5 to 11 x 11 x 4.5	200 - 500
Freshwater cage	5 x 6 x 4-6	120 - 180
Pontoon type steel cage	15 x 15 x 5	1125
'Viking' type steel cage	12 x 12 x 5, 15 x 15 x 5.6	720 - 1350
'Ewos' giant cage system	8 x 125square metres units x 8	8000
Bridgestone flexible cage	Hexagonal, 14 side x 10	5000 - 6000
Circular plastic pipe cage	5-20 diameter x 5-10	100 - 3000

[1] l = length
  b = breadth
  d = depth

Table 5.6:   Outline capital costs: cage salmon farm

	Annual Production		
	50 tonnes	200 tonnes	500 tonnes
Site preparation	500	3,000	5,000
Services, access	-[1]	2,000	4,000
Cages, moorings	20,000	80,000	180,000[2]
Nets	7,500	30,000	65,000[2]
Boats	1,000[1]	20,000	40,000
Vehicle(s)	-[1]	16,000	24,000
Equipment, feeders	-[1]	3,000	7,000
Store/workshops	-[1]	25,000	50,000
Accommodation	-[1]	35,000	60,000
Miscellaneous	2,000	5,000	10,000
Contingencies (10%)	3,100	21,100	42,700
Total	34,100	243,100	493,900

1   Provided by operator
2   Cages and nets can be used more efficiently.

Site preparation, services, etc., depend greatly on local conditions.

In some areas accommodation and/or storage can be leased.

The 500t unit assumes a minimum of 2 sites, serviced from a single main centre, plus subsidiary centre.

Table 5.7: Capital costs: more sophisticated systems

	Annual Production	
	200 tonnes	500 tonnes
	£	£
Site preparation[1]	5,000	8,000
Services, access[1]	4,000	8,000
Cages, moorings[2]	120,000	270,000
Nets	30,000	65,000
Boats[3]	25,000	50,000
Vehicle(s)	16,000	24,000
Handling equipment[4]	15,000	40,000
Feeders[5]	50,000	100,000
Store/workshops	25,000	50,000
Accommodation	35,000	60,000
Miscellaneous	8,000	20,000
Contingencies	33,300	69,500
Total	366,300	764,500

Notes:

1 Additional preparation for handling equipment, bulk deliveries, etc.

2 Steel cages - butyl cages approx. 30-50% more expensive.

3 Additional handling facilities includes purpose-built service boat.

4 Forklift, overhead gantries, hydraulic arms and winches.

5 Moist feed manufacturing unit with manifold and water-based distribution system.

suitable quality has led to an increased interest in using more exposed waters, by employing more durable cage systems. Although these systems are more expensive, there may be advantages of better water exchange and cleaner environments. More expensive and sophisticated systems are also being used in existing sites, particularly if improvements in handling and use of manpower can be demonstrated. Table 5.7 illustrates the capital cost implications of these developments.

Although enclosures have been used from the outset of commercial salmon culture (see e.g. Milne, (7)), their use has not generally developed to the same extent as that of cages. Table 5.8 shows a typical capital cost structure, but it must be emphasised that costs are extremely site-specific. Although, in favourable circumstances (e.g. sites with two or three sides already formed - as in bays or narrows), cost per unit volume can be substantially lower than cages, stock densities are also lower, management of stock is much more difficult, and removal of wastes may necessitate the use of aeration and/or propeller pumps. Furthermore, several operators report poorer growth performance and increased

Table 5.8:   Enclosures, capital cost structure

	Annual production		
	50 tonnes	200 tonnes	500 tonnes
	£	£	£
Site preparation	500	3,000	5,000
Services, access	-[1]	2,000	4,000
Enclosure structure[2]	16,000	38,000	80,000
Nets	2,400	6,000	12,000
Boat(s)	1,000[1]	10,000[3]	20,000[3]
Vehicle(s)	-[1]	16,000	24,000
Equipment, feeders	-[1]	3,000	7,000
Store, workshops	-[1]	25,000	50,000
Accommodation	-[1]	35,000	60,000
Miscellaneous	2,000	5,000	10,000
Contingencies	2,190	14,300	27,200
Total	24,090	157,300	299,200

1   Provided by operator
2   Assume 30% of total border length, 70% formed
    by land edge, 4 metres deep
3   Assume near land access. Remote site requires
    more boat capacity.

disease incidence in existing systems (Koltveit, personal communication).

Pumped seawater systems have been developed as a land-based alternative to cage production, theoretically offering more controlled conditions, removal of wastes from the immediate vicinity, controlled feeding, freedom from exposure problems, and simplified management and security. Capital cost requirements are significantly higher than those for cage culture, but they may be somewhat offset by improved lifespan (see Table 5.9). Costs are extremely site-specific, with the result that the costs of water abstraction and supply may represent a significant part of capital investment. Economies of scale are difficult to define at present, though the potential does exist, both in increased project size and increased tank size.

5.6 OPERATING COSTS OF PRODUCTION

5.6.1   Freshwater production

Table 5.10 shows typical operating costs for the freshwater stages of Atlantic and Pacific salmon production, and demonstrates the effects of different production methods and production strategies. It can be seen that at lower production levels, management and labour are the most significant inputs; as output increases, feed and stock costs become more significant. At present, production costs are significantly lower than market price (Table 5.11) reflecting the current market demand. However, some recent projects, some including a substantial premium for suitable land areas, are demonstrating potential production costs of up to 80-90% of current market price. Thus a trend may be observed of increasing capital costs in overall production costs.

5.6.2   Seawater production

The operating cost profiles for seawater production in cages, enclosures or tanks, are shown in Table 5.12 for a typical small scale production unit (50 tonnes) and for larger scale units (200t). These demonstrate the overall significance of stock and feed costs, and the relatively high management and labour cost of smaller commercial units. It is worth commenting at this stage that the different ownership

Table 5.9: Pumped seawater systems, capital costs

| | Annual production | | |
	50 tonnes	200 tonnes	500 tonnes
	£	£	£
Site preparation	5,000	20,000	50,000
Services, access	2,000	5,000	10,000
Tanks[1]	40,000	140,000	300,000
Water supply system[2]	7,000	50,000	120,000
Pumps[3]	8,000	30,000	70,000
Boats	-[4]	1,500	1,500
Vehicles	-[4]	16,000	24,000
Equipment, feeders	-[4]	4,000	10,000
Store/workshops	-[4]	25,000	50,000
Accommodation	-[4]	35,000	60,000
Miscellaneous	2,000	5,000	10,000
Contingencies (10%)	6,400	33,150	70,550
Total	70,400	364,650	776,050

Notes

1   8 and 12 metre diameter tanks at 50 tonnes, 12 and 18 metre tanks at 200 tonnes 18 30 metre tanks at 500 tonnes, based on 20 kilogrammes per cubic metre

2   Intake, stilling basis and water supply channel

3   Based on 5m pumping head, with average 0.8 litres per minute per kilo weight

4   Supplied by operator

Some recent projects have been initiated using higher stock densities and lower overall water flows. Their efficacy, given higher disease risk, has yet to be confirmed.

Table 5.10: Operating costs freshwater stage – Atlantic and Pacific salmon showing the effects of different strategies

£ sterling

Production/yr Numbers of fish	A			B	
	20,000	100,000	200,000	1,000,000	5,000,000
	£	£	£	£	£
Variable Costs					
Eggs[1]	1,000	5,000	10,000	15,000	75,000
Food	800	4,000	8,000	6,000[2]	30,000[2]
Insurance	800	4,000	8,000	8,000	40,000
Fixed/Semi-variable Costs					
Management	-[3]	20,000[4]	20,000[4]	20,000[4]	32,000[4]
Labour	-[3]	9,000[4]	18,000[4]	24,000[4]	40,000
Power/fuel	500	1,000	3,000	10,000	30,000
Consumables	300	1,000	2,000	2,000	8,000
Administration	300	600	900	3,000	12,000
Maintenance	300	1,000	1,500	1,500	6,000
Contingencies	400	3,960	6,240	8,950	27,300
Financial charges					
Working capital[5]	220	2,180	3,430	4,930	15,020
Fixed capital[6]	4,130	29,290	45,100	36,030	95,430
Total Costs	8,750	82,030	126,170	139,400	410,750
Unit cost	0.438	0.820	0.631	0.139	0.082
Estimated effect of					
use of freshwater cages[7]	0.400	0.720	0.550	0.120	0.075
use of recycle systems[8]	0.550	0.900	0.700	0.160	0.095

Notes
A   Smolt production, normally Atlantic, also possible for coho and chinook
B   Fry production, normally all Pacific
1   Assume purchased. Costs from own broodstock typically 30-70% of this. Based on 80% survival (A), 90% survival (B)
2   Fed to 8 grammes
3   Supplied by operator.
4   May be up to 100% higher in Scandinavia, North America
5   Based on 50% of annual operating cost, 10% interest charge
6   25% of fixed capital to cover interest/depreciation
7   Lower capital charges
8   Higher capital charges, power costs

Table 5.11:   Production costs compared with market
prices, Atlantic and Pacific salmon

	Selling price per 1000	Production cost per 1000
Atlantic salmon		
Eggs	£ 25 -£ 50	£ 5 -£ 20
Fry	£ 100 -£ 400	£ 50 -£ 150
S1 smolts	£1200 -£1800	£400 -£1200
S2 smolts	£1300 -£1800	£500 -£1200
Pacific salmon		
Fry	£ 150 -£ 300	£ 50 -£ 150

structure in Scotland, Norway and Canada, results
in a significantly greater number of smaller
owner-operated farms in Norway, an intermediate
amount in Canada, and, apart from a recent
regional project, very few in Scotland.

For pumped seawater systems, the relatively
higher costs of capital charges and energy are
apparent. To date the potential saving of feed
and labour costs referred to earlier has yet to be
demonstrated. Recent developments in large scale
recycle systems may however reduce site
dependence, pump installation and water exchange
costs, and alter the cost profile more
favourably. It must be stressed however that
this is as yet at a very early and unproven stage
of development.

In the use of enclosures, operating costs are
very similar to those of cage systems, as capital
costs are relatively low, and pumping is not
normally required. However, as noted earlier,
actual costs are extremely site specific, and the
use of expensive enclosure walls, with pumped or
aerated water may lead to costs similar to those
of tank systems.

At present, the relatively low stock cost

Table 5.12: Operating costs, seawater production,
Atlantic and Pacific salmon

£ sterling

	Annual Production				
	50 tonnes	200 tonnes			500 tonnes
	Atlantic in cages	Atlantic in cages	Atlantic in tanks	Pacific in cages	Atlantic in cages
	£000s	£000s	£000s	£000s	£000s
Variable costs					
Stock[1]	32.5	130.0	130.0	30.0	325.0
Food	50.0	200.0	180.0	200.0	500.0
Insurance	5.0	20.0	20.0	20.0	50.0
Transport/pack	10.0	40.0	40.0	40.0	100.0
Fixed/semi variable					
Management	-[2]	20.0	25.0[3]	30.0[4]	40.0
Labour	-[2]	35.0	28.0	40.0[4]	60.0
Power/fuel	2.0	5.0	42.0[5]	5.0	10.0
Consumables	3.0	10.0	10.0	10.0	25.0
Administration	3.0	10.0	10.0	10.0	20.0
Maintenance	1.0	5.0	10.0	5.0	10.0
Contingencies,(5%)	5.3	23.7	24.7	19.5	57.0
Financial charges					
Working capital[6]	5.6	24.9	26.0	20.5	59.9
Fixed capital[7]	8.5	60.8	72.9	60.8	123.5
Total	126.0	584.5	618.7	490.8	1380.4
Unit cost					
(per tonne)	£2,520	£2,923	£3,094	£2,454	£2,761

1a) 80% survival, costs of purchase of stock: £1.30
    (Atlantic), £0.30 (Pacific) per fish,
  b) 2.5kg average production weight.
2 Owner-supplied.
3 Including engineer.
4 Allowances for higher wage rates in North America.
5 Average 160 kw @ £0.03 per kilowatt hour.
6 Based on 50% of annual operating cost, 10% interest
   charge.
7 Total 25%, covering interest and depreciation - 20% for
   tank system.

offers significant advantages to Pacific salmon producers; preliminary indications are that growth and other performance factors are otherwise similar to those of Atlantic salmon. This is therefore of considerable significance for the future and if harvest prices of farmed coho and chinook match those of Atlantic salmon, there could be a considerable competitive advantage in favour of Pacific farmed salmon.

Apart from capital service costs, other operating costs are generally more standardised through the industry, as inputs are similar, and are either set (e.g. smolts) or competitive (e.g. feed) prices. The overall competitive position of those firms capable of producing their own stocks and/or using lower cost feeds (e.g. trash fish and silage based feeds), may therefore be stronger, though only recently has competition become significant. The main areas of difference in operating costs between producers lie in food conversion, mortality rate, and growth; Table 5.13 illustrates the effects of changes in these variables.

5.7 SYSTEM AND SCALE EFFECTS

The effects of system choice have already been described to some extent in the previous section but Table 5.14 summarises the main effects observed in both capital and operating costs. However, as already noted, the effects of site and development cost may still be significant within this, particularly with tank and enclosure culture systems, which are far less standardised.

Scale effects, apart from the specific features already discussed, are rather hard to establish, though in certain circumstances they may be present. In practice, variation in site characteristics, and the limitations of single sites, requiring multiple units, tend to dominate comparative evaluations and if such effects exist, they are relatively unimportant.

Thus it is possible for smaller operators, particularly family operated units, to be competitive with larger units, and in cases where particularly good growth conditions are present, profitability can be significantly better regardless of size of system. At present there are few means of objective assessment of site potential and there is little mechanism for

Table 5.13: Effects of efficiency on operating cost of production

£ sterling

	Annual Production				
	50 tonnes	200 tonnes			500 tonnes
	Atlantic in cages	Atlantic in cages	Atlantic in tanks	Pacific in cages	Atlantic in cages
	£000s	£000s	£000s	£000s	£000s
Variable costs					
Stock[1]	25.5	102.0	102.0	23.5	255.0
Food[2]	40.0	160.0	144.0	160.0	400.0
Insurance	5.0	20.0	20.0	20.0	50.0
Transport/pack[3]	9.0	36.0	36.0	36.0	90.0
Fixed/semi variable					
Management	–	20.0	25.0	30.0	40.0
Labour	–	35.0	28.0	40.0	60.0
Power/fuel	2.0	5.0	26.0[4]	5.0	10.0
Consumables	3.0	10.0	10.0	10.0	25.0
Administration	3.0	10.0	10.0	10.0	20.0
Maintenance	1.0	5.0	10.0	5.0	10.0
Contingencies (3%)	2.7	12.1	12.3	10.2	28.8
Financial charges					
Working capital	4.6	20.8	21.2	17.5	49.4
Fixed capital[5]	7.7	54.7	65.6	54.7	111.2
Total	103.5	490.6	510.1	421.9	1149.4
Unit cost					
(per tonne)	£2,070	£2,453	£2,550	£2,110	£2,299

1a) 80% survival, costs of purchase of stock: £1.30
    (Atlantic), £0.30 (Pacific) per fish,
  b) 3.0 kilogrammes average production weight.
2   Food conversion improved by 20%.
3   Reduced 10%.
4   Minimum possible head.
5   Reduced 10% to allow for higher stock density.

Table 5.14: Effects of system on capital and operating costs for Atlantic salmon

Hatchery: 100,000 smolts[1]
                    Capital costs   Total operating costs

    Tank   117.2   82,030
    Cage   96.5   72,000
    Recycle   131.5   90,000

Ongrowing: 200 tonnes production[1]
                    Capital costs   Total operating costs
Cage system
    standard spec.   243.1   584.5
    high spec.   366.3   610.0[1]
Enclosures   157.3   580.0[1]
Onshore tanks   364.7   618.7
Recycle system   380.0[2]   610.0[2]

Notes:
1   Differences due to sites and individual efficiency - may
    change any of these by 10-30% or more.
2   Estimate on the basis of capital charges, efficiency
    differences.

permitting competition for the use of sites according to production potential, particularly as in most cases a licence system restricts use and ownership.

Because of the control of site area and ownership, there is little evidence as yet of significant concentration of production in the industry, and though examples exist in Norway, Scotland and Canada of operators with large and/or multiple sites, there is no evidence of significant production cost advantages over efficient smaller producers. For existing large

producers there may however be a significant
incentive for increased concentration to reduce
the overhead costs of professional management
resources and because of the marketing advantages
of larger size.

5.8 RETURNS, RISK AND INVESTMENT

Like many forms of aquaculture production,
particularly in temperate waters with fish of a
long life cycle, cash flow patterns are
characterised by a relatively long period before
positive balances are achieved. However for
hatchery production alone without the on-growing
stage, particularly at present sales price levels,
cash flow becomes positive far more quickly (Table
5.15).
    There is thus generally a considerable period
of uncertainty, and in privately-run operations,
an outlay of fixed and working capital to which
substantial risk is attached. While overall
average returns to the operation may be good (see
Table 5.16), the year-to-year variation in
performances and the generally high level of
uncertainty associated with initial years of
production, e.g. site performance, relative
inexperience of operators, may make initial
development slower and more difficult than
expected.
    For private operators, the slow movement to
positive cash flow means that unless external
sources of finance are available and willing to
invest in operations not as yet fully proven to be
reliable, the generation of self-financing
capital, and hence the potential for expansion can
be correspondingly restricted. Returns to
hatchery operations are, for the moment,
substantially better than those for ongrowing
farms. This may however be temporary. While it
is clear that both sections are interdependent and
so some equalisation of transfer prices might be
expected, the importance to ongrowing farms of an
assured stock supply given the present rate of
expansion of the industry and the potential for
loss of freshwater stock through technical risk
has allowed smolt prices to stay at higher levels
than their rearing cost would indicate. The
importance of supply may be particularly
significant to those producers who are sensitive
to under-capacity (e.g. the enclosure and pumped

Table 5.15:   Cash flow patterns - hatchery and
ongrowing

Years	0	1	2	3	4
Hatchery, 100,000 smolts/year					
System	40.0	40.0	20.0	–	–
Variable costs	5.0	13.0	13.0	13.0	13.0
Fixed/semi variable costs	15.0	33.0	36.0	36.0	36.0
Total costs	60.0	86.0	69.0	49.0	49.0
Sales revenue[1]	–	–	78.0	130.0	130.0
Gross profit	(60.0)	(86.0)	9.0	81.0	81.0
Cumulative gross profit	(60.0)	(146.0)	(137.0)	(56.0)	25.0
Ongrowing, 200t Atlantic salmon					
System	80.0	120.0	40.0	–	–
Variable costs	260.0	330.0	390.0	390.0	390.0
Fixed/semi variable costs	60.0	80.0	110.0	110.0	110.0
Total costs	400.0	530.0	540.0	500.0	500.0
Sales revenue[2]	–	280.0	700.0	700.0	700.0
Gross profit	(400.0)	(250.0)	160.0	200.0	200.0
Cumulative gross profit	(400.0)	(650.0)	(490.0)	(290.0)	(90.0)

1 At £1.30 each
2 At £3.50/kg

Table 5.16:   Overall returns, salmon farming

	Returns, excluding financial charges, and tax		
	On operating costs	On capital	
	%	%	
Operation			
Atlantic salmon			
Hatchery, 20,000	491	131	1
100,000	157	68	
Pacific salmon			
Hatchery, 100,000	400	130	2
1,000,000	205	140	
Atlantic salmon			
50t cages	56	185	3
200t cages	40	83	
500t cages	46	112	
200t tanks	35	49	
200t enclosures	40	127	
Pacific salmon			
50t cages	71	182	4
200t cages	47	78	

1   Based on £1.30 sale price
2   Sale price £0.30
3   Sale price £3.50/kg
4   Sale price £3.00/kg

tank operations, with relatively high fixed
operating costs), who have a strong incentive to
stock even at high prices.
    In the longer run, the question must arise of
the break-even production price of the most
efficient producers in the industry. Table 5.17
shows the overall order of potential production
cost. While current sales prices have not
exerted a large downward pressure on profits as
yet, the increasing capacity of the industry may
well create such circumstances, and there are
already trends towards lower price levels.

Table 5.17: Overall levels of production cost

                        Lower
Hatchery: Costs
     Pacific salmon   Atlantic salmon
     Cage-based
     Tank systems
     Recycle systems
                                 Cage-based
                                 Tank systems
                                 Recycle systems

Ongrowing:
     Enclosure
     Cage, standard   Enclosure
     Cage, nigh spec.   Cage, standard
                                 Cage, high spec.

     Tanks, low head, recycle
                                 Tanks, low head, recycle
     Tanks, high head
                                 Tanks, high head.

                        Higher
                        Costs

These possibilities of market price constraints have encouraged a number of producers
either to co-operate in production and marketing,
or to develop vertically integrated operations,
based on the different quality and supply
characteristics of farmed salmon, in order to
secure a more stable position in the market (see
Chapter 7.)

At present, however, the risks associated with
salmon production are not perceived in general to
be excessive for the returns possible, investment
finance is in many cases available, and companies
from other areas are often interested in
diversifying into salmon farming. This is
particularly the case in Norway, where the
excellent general performance of the aquaculture
industry has given it a very favourable image for
investors. In the United Kingdom conditions are
rather more variable since although aquaculture is
perceived as a sector capable of growth and thus
potentially interesting for investment, long-term
risk capital is less readily available.

In Canada, investment, particularly in the
West coast, has been rather volatile, with
considerable speculative interest fuelled by
expectations of repeating the Norwegian industry's
performance, but little 'on the ground'
development has yet taken place. However, once
the initial developments indicate that production
is financially viable, it is likely that
substantial funds will become available for the
development of the industry.

It is interesting to note that the trend of
investment and build-up of production in the
original areas of development in Norway and
Scotland has to some extent been replaced, apart
from consolidation of existing operations, by
investment in outside areas where site potential
appears better and some markets may be more easily
served. There is thus a considerable investment
by Norwegian interests in Canada, and now in the
Southern Hemisphere.

One increasingly evident trend is the
commitment to investment by the aquaculture
service industry. A simple capacity matrix
(Table 5.18) demonstrates that sizeable supply
potential is now building up, particularly in
construction work (hatcheries) tanks, cages, and
feed, and many of these manufacturers have been
sufficiently encouraged by the growth in the
industry to invest in design or product

Table 5.18:   Capacity   and   services   required   -
example   based   on   10,000   tonnes   per
year salmon production[1]

Item	Quantity	Cost per unit	Total
			£000 sterling
Inputs			
Eggs	70 million	£0.04	280
Smolts	50 million	£1.30	6,500
Food, dry	18,000 tonnes	£500 per tonne	9,000
moist	3,000 tonnes	£300 per tonne	900
Chemicals	(£30 per tonne of production)		300
Product value	10,000 tonnes	£3,500	35,000

Capital items	Item		Value Estimated	£mn	
	Installed	Replacement	New		Total
Hatchery troughs	0.1	0.005	0.01		0.015
tanks	1.0	0.05	0.10		0.15
cages	0.5	0.1	0.05		0.15
Ongrowing tanks	2.0	0.1	0.2		0.3
cages	4.0	0.8	0.4		1.2
nets	1.21	0.4	0.12		0.52
Pipes, etc.	0.6	0.03	0.06		0.09
Pumps	0.3	0.05	0.03		0.08
Feeders	0.2	0.04	0.02		0.06
Boats	0.8	0.16	0.08		0.34
Vehicles	0.6	0.15	0.06		0.21
Buildings	2.5	0.1*	0.25		0.35

Note:
1   Assuming annual growth rate of production of 10%.

development and production capacity.

The   rôle   of   government   development
institutions   in   supporting   salmon   aquaculture
during the earlier stages can be significant in
the development of the industry.   In Norway many
units were financed initially by restructuring
grants payable on taking fishing boats out of
service.   Infrastructure   investment   at   both
specific   levels   (support   for   producers,
organisation, government research) and general
levels (roads, bridges and ferries linking more
remote   areas)   has   also   been   significant.
Similarly, in Scotland, the Highlands and Islands
Development Board has done much at both the
infrastructure level and in providing direct grant
or loan support for initial production.   In many

instances the individual firms, once established, are able to attract outside financing or to finance expansion internally. At present, the industry is considered in many areas to be reaching the point where specific assistance in production is less important than strategic support for marketing, quality control and environmental and disease management.

The rôle of insurance in investment and risk management has been rather variable. While many producers insure stock against specific disease loss, in practice the amount of effective protection available often does not justify the costs of such insurance. In theory the system would protect against large-scale catastrophic loss, but current conditions of policy may well restrict operators within very limited areas of cover. However, the insurance industry has played an important rôle in defining acceptable operating and husbandry standards, and so has contributed to the reduction of risk generally.

5.9 CONSTRAINTS TO THE DEVELOPMENT OF PRODUCTION

The single most apparent constraint to production in the longer term is economic: as markets become saturated, prices will tend to fall, and to depress profitability.

Capital costs are also increasing because more sophisticated systems are being used and in some cases more difficult, higher cost sites are being developed. This will similarly tend to reduce profitability. What is difficult to predict, however, is the speed at which these constraints will operate, a question discussed further in Chapter 8.

The increasingly apparent limitations created by wastes from the farming operations themselves are forcing many producers to seek alternative sites, to reduce stocks in existing sites, or to operate multiple sites with a 'fallow' period on each site to allow environmental conditions to stabilise.

The site supply constraint applies both to freshwater and saltwater production, though in the former case the high product value is encouraging the development of relatively high-cost recycled water systems, or pumped, heated groundwater systems which may reduce site constraints. Similar developments might possibly occur in

ongrowing, although at present it is still more
attractive to develop using conventional cage
units, particularly in newer areas such as British
Columbia, than to enter the relatively expensive
development of what is currently a higher-cost
method of production.

To date, supply of smolts has also been a
constraint to the on-growing industry; this is
reflected in the current premium paid. However,
the sector is expanding, and there are few
technical constraints to long-term supply.

The supply of feeds of good quality has been a
constraint, and is still problematic in some of
the more recently developed areas, but is not
significantly limiting. It is unlikely, however,
that feed costs will move substantially downward,
as these are based on prices of feedstuff
commodities which are sold on for larger markets
than that of the salmon industry. There may
however be improvements in feed utilisation, and
current research may also widen the range of
potential raw materials, and so tend to stabilise
prices.

The problems of disease and its control
remain; environment-related disease conditions
are likely to be of increasing significance as
site constraints increase. The development and
control of new diseases or the transfer of
diseases from other areas continues to be a
significant threat to the industry. Although to
date the industry has generally been able to
control losses with specific management measures,
the severity of individual losses can be extreme
and if management measures are not available,
could imply substantial loss for the whole
industry.

The supply of skilled staff has until
relatively recently been a constraint, but the
availability of training, and the build-up of
staff resources within the industry have reduced
this problem. However, supply of higher level
management may still be a local constraint.

5.10 SUMMARY

The salmon aquaculture industry is now a major
component in the structure of salmon supply. The
potential for year-round production of relatively
uniform quality, with a range of sizes, has
permitted significant changes in the way in which

consumer demand is met. With a relatively well-developed technology, and as yet a strong market demand, the industry has grown rapidly, and in fact may appear to be growing in a 'production-led' rather than a 'demand-led' manner.

The potential for additional production at similar levels of development cost as at present is substantial; in British Columbia a potential of 50,000 tonnes per year by 1990 has been mentioned by some commentators. The levels of production involved will have a substantial effect on supply in this sector, which will have to be met either by market expansion or by clearing at a lower price level.

The development of new technologies, particularly in stock improvement, control over reproduction (to limit losses of input feed), new tank systems, and waste-removal devices, will facilitate development, but may not lower production costs substantially. Although larger-scale systems could potentially reduce overall production cost, the development risks are also substantial, and overall disease risk may also create constraints. Thus the longer term development of salmon aquaculture will have to be based on the existence of a market size sufficient to support prices at around or near the present production cost levels.

REFERENCES

1   D J Edwards, Salmon and Trout Farming in
    Norway, 1978, Fishing News Books Limited,
    Farnham
2   A M Sutterlin and S P Merrill, Norwegian
    Salmonid Farming, April 1978, Fisheries and
    Marine Service Technical Report No. 779,
    Government of Canada
3   T Mowinckel and H Kvalheim, Atlantic Salmon
    with special reference in enclosure systems,
    2nd European Fish Farming Congress 1976, London
4   J   E   Thorpe   (ed),   Salmon   Ranching,   1980,
    Academic Press, London and New York
5   J E Muir, Management and cost implications of
    recirculating water systems, Proceedings of
    the   Bio-Engineering   Symposium   for   Fish
    Culture, 1979, American Fisheries Society and
    the Northeast Society of Conservation Engineers
6   G   Boeuf   and   Y   Harache,   Present   status   of
    salmonid cage rearing in Western France and
    prospects for development, EIFAC
7   P H Milne, Fish and shellfish farming in
    coastal waters, 1972, Fishing News Books,
    Farnham

FURTHER READING

An Bord Iascaigh Mhara, The Atlantic Salmon
    Farming Industry, Past Performance and Future
    Potential, 1986, BIM Market Research Series,
    Dublin.
The Institute of Fisheries Management, Proceedings
    of the 15th Annual Study Course, 1984,
    Stirling.

Chapter Six

PROCESSING SALMON

6.1 INTRODUCTION

After harvesting a number of activities take place
to match harvesting patterns with the requirements
of customers.   These are sometimes carried out by
the producers themselves and in other cases they
are   carried   out   by   separate   processing
companies.   In each case these activities involve
processing the salmon, transporting, storing and
sorting   the   product   and   communicating   with
customers   so   that   their   requirements   are
understood and transactions with them can take
place.   Communications   are   either   by   direct
personal contact or through advertising and other
forms of promotion.   During this stage the value
of the product more than doubles (Table 6.1) so
the stage is of considerable economic significance
and its performance has a major effect on the
well-being of the industry as a whole.
        The   main   forms   of   processing   are   the
preparation of fresh fish, freezing, canning and
smoking.   In   the   sections   below   the
characteristics of each type of processing are
outlined, the structure of the processing sectors
is   described   and   an   explanation   provided   for
patterns   of   organisation   in   the   processing
industries.   The chapter draws heavily on the
North American experience because this is the best
documented.

6.2 FRESH AND FROZEN SALMON

6.2.1   Processing
From the moment of slaughter, the fish undergoes
physiological and biochemical changes which will

Table 6.1:   Index of value added in processing
             Alaskan salmon

	US sockeye 1/4 can sold in US	Frozen sockeye sold in US	Frozen coho sold in US
Ex vessel round	100	100	100
Converted to processed weight	143	133	133
Processor selling price	430	270	320
Smoker selling price	-	-	-640

Source: Industry estimates
Note: These are indicative only.   There are
          fluctuations over time, between species
          and quality of fish.

have significant effects on the final product
quality.   Initially relaxed, the fish muscles
stiffen several hours after slaughter, during the
'rigor' stage, after which the muscles again
relax.   From this point onwards, unless
specifically treated, the tissues deteriorate
progressively and quality diminishes.   Bacterial
and enzyme action, particularly from the gut,
cause changes in flesh quality and texture, which
occur rapidly at higher temperatures.   Good
initial handling, minimising stress to avoid
lactic acid accumulation in the tissues and
consequent poor keeping, and avoidance of bruising
and flesh discolouration, together with cooling in
the pre-rigor stages, is particularly important in
basic quality control.
    Fish should be slaughtered quickly and cleanly
and with the minimum of handling stress to avoid
lactic acid accumulation in the flesh, as this
results in poor keeping qualities.   Farmed fish
should be starved for at least three days prior to
slaughter.   Fish may be bled at slaughter to
improve keeping qualities, a process which is
normally associated with farmed or ranched salmon
where the appropriate facilities can more easily
be provided at the point of slaughter.   Fresh
salmon are sold whole, but they can also be gilled
and gutted.   Frozen salmon, particularly from
North America, are usually dressed, i.e. headed as
well as gilled and gutted.   The weight loss for

gilling and gutting is around 8% and the weight loss for dressing is 15-18%.

Fish for fresh sale should ideally be iced within a maximum of 60 minutes of slaughter, using flake ice (at least 20% of fish weight) or alternatively chilled in an ice/brine mix. Fish should be packed carefully to avoid twisting or compression. Iced fish should ideally reach the final point of sale within 72 hours of slaughter if they are whole and 96 hours if they are gilled and gutted.

Fish for freezing can be frozen whole, gilled and gutted or dressed. They should first be chilled then blast frozen as single fish for at least 12 hours. They should be frozen prior to or post rigor to reduce "gaping" on thawing. The freezing temperature is normally less than $-30^{o}C$ and preferably $-40^{o}C$. After this the fish can be dipped in a glazing solution of water and fructose which seals the fish and improves the appearance. Alternatively they can be contained in moisture proof material for the same purpose. The fish should be stored at $-20^{o}C$ or preferably $-30^{o}C$. They should not be thawed and re-frozen prior to sale. Frozen salmon can be stored without deterioration of quality for 6 months to one year.

Premium quality fish should show negligible scale loss, no external signs of deformity, maturation damage or lesions and there should be minimal erosion of fins and tail. They should have a clean fresh smell.

The fish are usually transported in boxes of 20 to 25 kg, marked with the name of the supplier, the date, the weight and the number of fish. For fresh fish, boxes should be insulated for all but the shortest distances. Fresh fish which are air-freighted travel in boxes with an absorbent lining for the melt water. Fresh and frozen salmon are normally sold by weight category of the individual fish. For instance Atlantic salmon are normally sold in the following size ranges: 1-2 kilos, 2-3 kilos, 3-4 kilos, 4-5 kilos, 5-6 kilos and above 6 kilos. These differences are significant since fish of different weights often go to different end uses.

Because these operations are largely manual and have to be carried out quickly and carefully after harvesting, they are often carried out by the fishermen and fish farmers themselves unless distances to processors are short. This is the case with some farmed fish, troll caught fish and

153

fish caught with coastal nets. With gill net fish and seine net fish the quantities caught at any one time are too large for the fishermen to handle so the fish are delivered to shore based processing plants or to tender vessels.

If Pacific salmon are to be frozen this is usually carried out on board ship; for example, this is the case in the Japanese high seas fishery and on some troll boats, or they are delivered to tender vessels and shore based processing plants. In the latter case, freezing is carried out in plants owned by the processors or in public cold stores where the fish can be frozen and stored on a contract basis. The small quantities of Atlantic salmon which are frozen are usually handled at fish processing plants or at contract cold stores.

6.2.2 Marketing

The marketing function requires decisions about the markets towards which the products are to be targeted, the types of processing required by those markets, pricing, choices of the distribution system to be used and decisions on methods of communication with customers.

Processors of fresh and frozen wild salmon have strategies concerning the general markets they are seeking to serve and they will attempt to develop these markets through their marketing activities. These decisions will depend not only on the state of different markets but on the characteristics of the supplies available to the processor by volume, species and quality as well as the transport and distribution options available to them to reach the different markets. Often however these decisions cannot be taken in advance of the season but have to be taken on an on-going basis during it because of variations in catches and the difficulty in predicting catch levels and qualities in advance. Price plays a key rôle in determining product destinations and in indicating the current state of the market. For frozen salmon the total level of stocks held and their location also has a major impact on the state of markets, an effect which can involve complicated time lags because of the durability of stocks.

The sellers of farmed salmon by contrast have a potential marketing advantage over those dealing in wild salmon. This is because the quality and quantity of their product is predictable and controllable to a far greater extent with the result that sales and production can be planned in line with trends in demand. Further, because the quantities available can be predicted more accurately, it is possible to plan on a longer term basis which markets are to be served and how the products are to be marketed. Although so far these advantages have not always been realised, steady progress is being made in this direction.

Arrangements must be made to finance deals with customers. For international transactions, this includes assessment of actual and expected movements in international currency values. North American dealers usually quote selling prices in their own currencies and expect to be paid in those currencies. This may appear to reduce the risk to them caused by currency fluctuations, but this is illusory. Buyers in other countries are less willing to buy if their currencies depreciate and vice versa, which in turn affects the price in North American currencies. Futures markets for salmon are not well developed because the product is fairly perishable and prices fluctuate widely during the season. Dealing in international salmon markets is accordingly a highly skilled activity requiring specialist knowledge.

Distribution decisions are influenced by the size of the processors and the requirements of their customers. These issues are the subject of Chapter 7.

Concerning promotion decisions, the methods used by companies and the amount they spend on promotions depends on the company, the product and the market. However, in general most promotion by individual companies is directed towards other members of the marketing channel, i.e. towards wholesalers, retailers and caterers, rather than towards the final consumer. This is particularly true for fresh salmon. Fresh fish are normally sold "on the slab" and judged on the basis of appearance, price and the final seller's recommendation. The name of the original supplier is not relevant to final consumers. In any case since many processors are not supplying products on a year round basis it would be difficult to make the final consumer very conscious of the source from which fish come, so that it is difficult to build up brand identities for individual companies. This may however change somewhat in the future with farmed salmon because it is continuously available and at least one processor/supplier of farmed salmon already tags fish with a brand name to build up brand awareness. Troll fishermen are also beginning to tag their fish in order to build up a high quality image for the product. Another reason for concentrating promotional efforts in the marketing channel rather than at final consumer level is that much fresh and frozen salmon is sold to smokers and from that point onwards the product is identified with the smoker rather than the original source of the raw fish. Therefore it makes more sense to direct promotion towards the smokers who make the key decisions on choice of suppliers. Exceptions to this rule are those companies selling frozen or vacuum packed products. Here the pack itself can be clearly identified with the supplier because of the opportunities presented for attractive labelling. This means that advertising to the final consumer can more easily be related to the product itself and is consequently more worthwhile.

Expenditure on promotion by processors seems to vary widely. Some companies do little more than ensure that their names are in the lists of sellers published by magazines and trade associations, while others advertise heavily in trade magazines, participate actively in shows and in public relations work. Much depends on their customer profile. Troll fishing companies for example who often sell primarily to a small number of smokers prefer to communiate directly with customers through personal contact. On the other hand, large less specialised companies, particularly those selling a range of seafoods to many customers, are more likely to spend heavily on advertising and other forms of promotion because the large size of their customer network makes this the most cost effective method.

6.2.3   Industry Structures
6.2.3.1   The variety of patterns.
The companies involved in fresh and frozen salmon processing are very heterogeneous, reflecting both differences in their locations and in the market segments which they serve. They vary from companies which process and market fresh or frozen salmon as their sole business activity to companies with wide-ranging interests in salmon, in other seafoods and in other areas of food processing. They range from family owned small

firms to multinational companies operating in many product areas. Among the larger companies, there are several international link-ups. For example, some of the Japanese fishing companies own or partially own companies operating in North America. Recently some Norwegian trading and farming companies have been extending their trading interests into other European countries and into North America as part of an international interest in the production and marketing of farmed salmon.

Because of this diversity and because companies may not be competing directly with each other, generalisation about structural patterns and direct comparisons between companies are not easy. For instance, companies specialising in the processing of high quality troll fish may have quite different customer patterns from those processing large volumes of seine caught fish. From studies carried out, mainly in North America, it is however possible to provide some statistical description of structures and some analysis of the key factors affecting the size and structure of processing operations.

6.2.3.2. Structures.

Precise figures are not available but there are certainly large numbers of companies involved in processing fresh and frozen salmon. As an example, Table 6.2 gives an indication of the structure of the processing industries in Alaska and British Columbia. In Alaska in 1976, ignoring interties between companies, there were 105 companies with 124 plants processing and marketing 19,000 tonnes of salmon (1). In British Columbia in 1980 there were 105 firms handling around 30,000 tonnes of fresh and frozen salmon (2), as well as a number of fishermen and fishing companies carrying out their own marketing. Moving to Europe, in Norway in 1983 there were over 70 licensed exporters of around 18,000 tonnes of farmed salmon.

The levels of concentration in Alaska and British Columbia are relatively modest in comparison with many other manufacturing industries (3) and they are lower than the levels of concentration found in salmon canning which are discussed in the next section. However, since in North America the largest freezers of salmon are also the largest canners, the interrelationships

Table 6.2: Concentration in fresh and frozen salmon processing

	Number of firms	Number of plants	% output 4 largest firms	% output 8 largest firms
Alaska 1976	105	124	21	36
British Columbia 1980	41	n.a.	63	n.a

Sources: F Orth *et al*., <u>Market Structure of the Alaska Seafood Processing Industry, Vol. II Finfish</u>, University of Alaska Sea Grant Report 78-14 1981.

R Schwindt <u>Industrial Organisation of the Pacific Fisheries</u>, McDaniels Research Ltd, prepared for the Commission of Pacific Fisheries Policy Vancouver 1982.

of salmon interests have to be considered to give a full measure of the buying and marketing strengths of these companies, a point developed in section 6.3.

6.2.3.3 Production and transport costs.
The structure of fresh and frozen salmon processing can be explained by examining the structure of production costs, and the nature of marketing and other management activities.

The first of these issues, production costs, probably provides the main reason why levels of concentration are low. Firstly, the capital costs of entry into the industry are low compared with those of entry into many other industries. It is estimated that the costs of a freezing plant, with limited cold store capacity, capable of freezing 20 tonnes of salmon in a 24 hour cycle, are £300,000. For a plant with double the capacity they are £500,000. If operated for 8 hours a day such a plant could freeze 400 tonnes of salmon per month. These costs are low compared with the costs of acquiring new fishing vessels or fish farms. In any case it is possible to by-pass these capital costs altogether by the use of contract freezing facilities which are available in the main centres of salmon processing.

Secondly, cost structures are not such as to give a major advantage to larger firms since there are not major economies associated with production or transport as the size of unit is increased. It is estimated that (Table 6.3) production costs are not significantly decreased with production size. The degree of utilisation of plant of any

Table 6.3: Operating costs of freezing plants for a 90 day season

£000s

	20 tonnes per day capacity		40 tonnes per day capacity	
	Capacity 50%	Operation 100%	Capacity 50%	Operation 100%
Costs				
Labour	15,000	20,000	20,000	35,000
Power	20,000	35,000	40,000	70,000
Vehicle	2,000	3,000	3,500	5,000
Packaging	30,000	60,000	60,000	120,000
Maintenance & insurance	15,000	18,000	20,000	38,000
Contingencies	8,200	13,600	14,350	25,800
Total	90,200	149,600	157,850	283,800
Total pro- duction (tonnes)	900	1,800	1,800	3,600
Cost per tonne excluding interest charges (£)	100.2	83.1	87.7	78.8
Interest on capital expenditure at 12%	33,300	33,300	57,600	57,600
Cost per tonne including interest (£)	137.2	101.6	119.7	94.84

size is also critical in determining the actual
level of operating costs per unit of output.
    This conclusion is supported by the results of
other studies. Schwindt (2), quoting earlier
Canadian studies, suggests that by the time
interest costs on capital expenditure have been
taken into account, there is little difference
between the costs of the large processor in
British Columbia with his own facilities and the
costs of the smaller firm using public storage.
In any case (Table 6.4) direct costs are a small
percentage of total costs so that processing costs

Table 6.4:   Structure of processing costs for
              Alaskan salmon
                  % breakdown

	Sockeye frozen %	Sockeye canned %
Raw fish cost (round weight	50	44
Raw fish cost (converted)	67	63
Labour	6	7
Packaging and freight	8	9
Overheads, selling distribution	15	17
Finance	4	4
TOTAL	100	100

Source: industry discussions.

are a relatively unimportant component in selling
prices. For instance, skill in buying the
appropriate quality of fish at the lowest price
could outweigh any variations in processing costs
caused by different sizes of plant. Finally, as
was shown in Table 6.2, in Alaska most firms
operate more than one plant. This suggests that
it is more cost effective to operate a number of
smaller plants close to landing points than to
operate one bigger plant.
    There are transport savings associated with
increased load size but they are achieved at low
volumes of output. It costs, for instance,
approximately 40% less to transport 18 tonnes of
salmon (a full container load) from Bristol Bay,

Alaska, to Seattle than it does to transport half
that quantity. This represents a considerable
disincentive to ship smaller quantities
particularly for lower value species. But beyond
this point there are no major savings and given
the large volumes handled in North America, even
the smaller operators would not have difficulty in
handling full container loads.

Transport costs may be a more important issue
in salmon farming for the smaller farmers, some of
whom are producing less than 50 tonnes per annum
and are located in remote areas. In many cases,
however, this disadvantage can be offset by
collecting systems which bring the output of a
number of farmers to a central collection point.
For instance, on the Island of Lewis in Scotland,
a marketing organisation buys the output of
different farmers which it then ships off the
island in bulk quantities. In general, the
Scottish experience suggests that remote
processing plants suffer from cost disadvantages
in the early days when volumes of output are
small. As output grows the disadvantage lessens,
both because the size of loads to be transported
moves towards more economic levels and because of
the associated development of specialist contract
haulage systems which can assemble supplies from
different points.

6.2.3.4   Marketing issues.

It is normal for processors to spend money on
marketing activities, but, as Table 6.3 shows,
this is not a high percentage of total costs.
Therefore, even if some advertising economies of
scale do exist, such as discounts in bulk
advertising and economies in the administration of
larger volumes of sales, marketing costs appear
unlikely to give larger processors and traders a
major advantage.

Marketing activities may nevertheless provide
some explanation for those differences in size
that exist between businesses. One explanation
is that the larger firms have simply been more
successful in their marketing and have therefore
obtained larger market shares which they have
successfully retained. Taking the example of
Norway, it is estimated that 80% of Norwegian
farmed salmon is handled by 4 or 5 companies, out
of a total of over 70 licensed exporters. Larger
traders started developing markets for farmed
Atlantic salmon by using resources from their

other activities, which were often concerned with
the marketing of other types of fish. Part of
this development has involved liaison with fish
farmers to ensure that the salmon that they were
handling was of good quality and available on the
terms required by customers. As a result they
have built up reputations which have gained more
customers and have increased the volumes which
they handle. For these reasons, they retain the
loyalty of customers, provided that their prices
are competitive. Further, companies need to be
large and to market large volumes in order to
supply the needs of larger smokers and wholesalers
for particular product assortments and volumes.
So far, smaller Norwegian exporters are able to
compete with the larger exporters because of the
low capital costs of entry into salmon dealing,
but they have tended to deal either in spot
markets or in specialist market sectors where
competition is less direct. So far at least they
have not been able to pose a threat to the larger
and longer established companies. As long as the
larger exporters continue to offer competitive
terms to both suppliers and customers, it would
seem likely that this pattern will continue. The
survival of smaller companies is probably also
helped by the desires of downstream customers to
use many different sources of supply. This helps
to reduce the risks inherent in dependency on a
very small number of suppliers.

6.2.3.5 Market power and buying power.
The market power argument suggests that in general
larger firms get better selling and buying terms
because their size advantage confers bargaining
power. It is difficult to assess the relevance
of this argument for fresh and frozen salmon
because of the limited evidence available and
because of the interaction of marketing and
production issues. Nevertheless, the argument
does not seem likely to apply as far as sales
activities are concerned. Although international
markets are very segmented, there is a large
number of processors competing in most sectors and
no individual company has a dominant market
share. In other words, there are not many
selling situations where individual sellers are
able to dominate the market and get particularly
advantageous terms as a result.
    When processors buy salmon, however, some
commentators have suggested that they have

power and that as a result
et better terms than small
nt is supported by the fact
)cessors buying fish in each
ller than the number of
with each other to sell
n international markets. Once
documented evidence available
th America but since, as already
rgest freezing companies are also the
ners of salmon, this argument is
until the general economics of salmon
have been considered.

6.3 SALMON CANNING

6.3.1   The Process

Since almost all canned salmon is Pacific salmon
and the largest quantities come from the United
States and Canada, this section is concerned
exclusively with the North American canning
industry.

Fish for canning are mainly caught by the
gillnet and purse seine fleets. They are
delivered direct to the canneries or delivered to
tender vessels sent by the canneries to the
fishing grounds. The tender vessels carry
crushed ice or chilled seawater to preserve the
fish in transit to the canneries. Fish for
canning do not have to reach the same quality
standards as fresh or frozen salmon. Some
softness, scale loss and discolouration are
allowable but the canning process itself is the
most complex salmon processing method and has the
most demanding quality control standards. The
processing flow is shown in Figure 6.1.

Fish are delivered round to the cannery. The
roe are then extracted and processed separately.
Fish are then headed, split and gutted, cut into
portions and filled in cans. The cans are
weighed, the tops sealed and placed in a retort
for cooking. A 1/2 kilo can is cooked at a
temperature of 110°C for 90 minutes, while
smaller sizes require slightly less time. After
cooking, the cans are cooled, sheathed in plastic
wrapping and boxed. Sizes are in imperial
weights – 1/4 lb (180 grams) packed 48 to the
case, 7 and 3/4 ounces (218 grams) packed 48 to
the case, 1 lb (450 grams) packed 48 to the case
and 4 lb (1.8 kg) packed 12 to the case. In most

Figure 6.1:   Processing Flow Chart: Canning

CLEANING, GRADING (size, species, condition)

FRESH/FROZEN
PRODUCTS

HEADING, SLIMING,
FINNING,   SPLITTING,
GUTTING

KIDNEY REMOVAL

CUTTING

CAN FILLING

SALTING

CHECKING, WEIGHING

SEALING

COOKING IN RETORT

COOLING

BOXING AND WRAPPING

ONWARD SHIPMENT

plants equipment is highly automated and quality control is rigorous to avoid contamination of the cans. The weight loss on canning varies between species but is usually at least 30% from round fish to canned weight.
Canned salmon is stored unlabelled at the expense of the canner (or "packer") until sold. Prior to sale each consignment is examined and a product inspection is carried out (see Table 6.5).

Table 6.5:   Selling salmon

---

1. Offers of frozen Pacific salmon

    The offer would normally contain details of:
species	packaging
grade	glaze allowance
size of offer	source of processing
catch location	price
processing form	shipment dates
credit and payment	terms

2. Offers of canned Pacific salmon

    As well as details of species, size of offer, price, shipment dates and financial arrangements, there is a "cutting" report from the National Food Processors Association specifying:
    Colour (5 point grading scheme)
    details of number of cans (if any) showing:
    low weights, watermarking, mixed species, poor cleaning, low vacuums, bruises, poor filling.

6.3.2   The Marketing of Canned Salmon
Customers who buy unlabelled canned salmon from the canners buy on the basis of price, species and the results of the product inspection report.
Thus the main marketing effort of the canner involves trying to anticipate buyer requirements and negotiating final details of the sale, such as price and credit terms. From this point the name of the original canners is of little significance to wholesalers, retailers and final consumers. The canner's name can only be identified by

interpretation of a punched code on the can and
this code is unintelligible to final consumers and
to most retailers. The country of origin is
given some but not major prominence on the label.
Most canned salmon trading internationally is sold
in this way so that at this stage marketing
activities are somewhat similar to those for fresh
and frozen salmon.

Can labels carry the name of the importers,
wholesalers or large retailers in the market in
which they are to be sold. These labels are an
important part of the marketing process for canned
salmon. Because consumers can only judge the
contents of the can after purchase and not by
visual inspection at the point of sale, they use
these labels to indicate different qualities of
salmon. This means that consumers become very
loyal to the brands that represent the quality
that they want and this helps to maintain the
market position of existing companies. In the
United Kingdom, for instance, the same major
importers using the same brand names have
dominated the market for many years and it is
estimated that the largest three importers are
responsible for over 75% of sales (7). These
importers support their products strongly by
advertising to help maintain buyer loyalty.

In domestic markets in Canada and the United
States, the canners use their own brand names and
spend heavily on advertising and other forms of
promotion to help to build brand loyalty.
Because of this heavy investment in their own
domestic brand labels their supplies will be
committed to the domestic market first (2) and to
some extent export markets will be treated as a
residual. Often it is the larger canners who are
in this position. By contrast, smaller canners
who have not established such strong brand
positions on the domestic market will give export
markets priority.

6.3.3   Industry Structures And Production Costs

In North America, as Tables 6.6 and 6.7 show,
there are fewer firms and thus higher degrees of
concentration in salmon canning than in salmon
freezing.

In Alaska there is a moderately high level of
concentration since four firms market 46 per cent
of Alaskan output although it should be noted that
the level of concentration in some regions is much
higher than the average for the state as a

Table 6.6:   Structure of the salmon canning
              industry in Alaska 1973-5

	Number of firms	Number of plants	Total prod. (tonnes)	% output 4 largest firms	% output 8 largest firms
Regions:					
Southeast	14	15	7954	61	90
Central	21	28	16785	59	86
Western	13	14	7416	54	91
Arctic-Yukon Kuskokwim	46	63	32881	46	64

Source: F Orth *et al.*, op.cit.

Table 6.7:   Structure of salmon canning in British
              Columbia 1980

Number of firms:	115
Total production	24315 tonnes
Share of output of 4 largest firms	59

Source: Schwindt, op.cit.

whole.   In British Columbia 4 firms process 59%
of supplies.   In both areas the tables suggest
that concentration ratios for firms are higher
than that those for plants, so that concentration
of ownership is greater than that of production.
The larger salmon canners also freeze salmon in
substantial quantities so that the same firms
dominate both the canning and freezing sectors.
In British Columbia for instance the 11 largest
canners are responsible for 80% of the output of
frozen salmon (5).   In part this is because
expansion into freezing was a logical direction
for diversification by the canners as the markets

for frozen salmon started to grow. In part it is also because of the benefits of being in both sectors of the market, as this gives flexibility in handling different grades and species.

Explanations for the structure of the canned sector can be found by looking at the same issues that were examined for frozen salmon: costs, marketing and market power. As far as costs are concerned, it has been argued that production economies of scale and initial capital costs are higher in canning than in freezing (6) and provide a barrier to the entry of new small firms. In 1984 the estimated capital cost of one canning line was US £540,000 and the cost of leasing a canning line was about US £54,000 for the first 2 years, declining subsequently. As far as economies of scale are concerned, the UMA/ERA report (8) suggested that canneries with a single line were at a 40% cost disadvantage compared with a plant with 12 to 15 lines. Canneries with 5 to 6 lines were at a 15% cost disadvantage compared with the larger plants.

Subsequent analysis has however suggested that barriers are less significant than had been earlier thought. Firstly there is not unanimous agreement on the validity of the figures presented above. Pinkerton (5) has suggested that the well organised small canner does not suffer such a large cost disadvantage. The way in which the flow of fish is organised and the way in which labour is organised, she has argued, are more important in determining costs than the number of canning lines operated. Secondly, plant concentration ratios are lower than firm concentration ratios (Tables 6.6 and 6.7), i.e. the larger firms operate more than one cannery. This means that the size of firms is not explained by the structure of costs at plant level. There is also no evidence of economies of multiple plant operations (2).

Finally, the major item in costs is raw fish (Table 6.3) and thus the buying price for raw materials could have a more important effect on competitiveness than the number of canning lines. This conclusion is supported by the successful entry of a number of new small firms into salmon canning in the 1970s, all with less than 4 canning lines. It appears therefore that the structure of production costs cannot completely explain the structure of firms in this sector.

6.3.4   Marketing Costs
Marketing costs help to explain why domestic markets are dominated by larger firms, but are less helpful in explaining patterns in export markets. In domestic markets it was noted that the large canners market under their own brand names and maintain brand loyalty through heavy promotion. Smaller firms attempting to challenge the position of these dominant firms can find the advertising outpay necessary to compete prohibitive. This barrier may have been reduced somewhat recently because more retailers are buying unlabelled salmon and using their own brand names but the number of successful new entrants into domestic markets has nevertheless been small. By contrast in export markets, the smaller firms do not face such a barrier because most salmon is sold unlabelled. Thus marketing costs appear to affect the choice of market rather than the ability of smaller canners to compete.

6.3.5   Market Power
There are two issues here: market power in selling processed salmon and market power in buying raw fish.
On the sales side none of the existing processors are large enough to dominate international markets by virtue of their size. There are approximately 11 firms canning salmon in British Columbia, over 40 in Alaska as well as the canners in Washington State, Japan and the USSR. Given that in recent years canned markets have been sluggish, buyers could readily turn to alternative sources of supply if they did not like the terms offered by individual canners. Thus size does not appear to have been a critical factor.
On the purchasing side there may be more substance to the argument that large processors have buying advantages because of their size. Typically, larger processors operate numbers of collecting stations and offer non-price services to fishermen such as credit, pre-season loans and repair services for boats. These services are valued by many fishermen who continue to sell to the large processors even if prices offered by other processors are higher. Accordingly, in many cases the decision to supply a large buyer is near automatic.
In practice, also, although there may be a large number of buyers in total, there may not be

many buyers at the particular landing points used
by the fishermen, so that concentration of
regional buying power by the large buyers is
probably greater than the overall figures
suggest. For instance, in British Columbia alone
one processor accounts for nearly 40% of all
purchases of raw salmon.

Whether this situation has actually given the
larger buyer substantial power is more difficult
to say. The most detailed studies have been of
the relationship between processors and fishermen
in British Columbia and different studies have
produced opposing conclusions. Schwindt (2) has
suggested that the larger firms do not have
substantial power, quoting the evidence that
actual prices to fishermen have been continually
above the minimum price levels set at the
beginning of the season by the fishermen's
unions. This would indicate that the larger
firms have not been acting as a joint informal
buying group to keep prices down. The fact that
the fishermen's union exists and does negotiate
prices also suggests that fishermen have some
countervailing power against the large
processors. Pinkerton, by contrast, has
suggested that the fact that fishermen are
dependent on processors for a range of services
means that many are forced to sell to the large
processors even if they could get higher prices
elsewhere. Because of these conflicting findings
it is therefore difficult to be conclusive on this
issue, except to note that it is clear that
conditions operating in the markets for raw fish
in particular areas can be very different from
those in the international markets in which the
processors sell.

6.4 RELATIONSHIPS BETWEEN FISHING, FARMING AND PROCESSING

6.4.1   Backward Integration Into Fishing

So far the analysis has been concerned with the
number of firms engaged in processing, while
ignoring the fact that not all processors and
harvesters are separate from each other and that
in fact linkages exist between the two. It is to
this issue of integration between stages that we
now turn, but since so many processors combine
canning and freezing, this section does not make
any distinction between the two activities.

170

Turning firstly to fishing and processing,
backward integration, i.e. the ownership of
fishing boats by processors, exists in both North
America and in Japan. In Alaska 9 out of 13
respondents in a 1976 study of processors owned
fishing boats and in one case a processor owned
over 30 boats (1). Such tie-ups are however far
from universal and have become of decreasing
significance in recent years as a result of the
establishment of the limited entry fishery.
Since this gave economic rights to fishermen by
giving them valuable and marketable rights to
fish, fishing activities have become less
dependent on capital from processors. In British
Columbia the pattern is similar. In 1979
processors owned a quarter of the seine fleet, 18%
of the gillnet boats but none of the troll boats
(whose fish do not go to canning). In Japan,
however, the high seas salmon boats are owned by
large fishing companies who are heavily engaged in
processing and marketing the fish of many
different species.
    In seeking explanations of why such
integration might take place, three issues may be
considered relevant: production costs,
transactions costs and market power. It is
difficult however to see the first of these,
production costs, as an important factor in
backward integration. Fishing and processing are
quite separate activities which, with the
exception of high seas fisheries, cannot be
combined into a continuous joint process. The
typical processing plant operates by combining the
inputs of many different boats arriving at
different times so that plant can be run
continuously. An efficient plant can absorb the
output of many boats, particularly in canning, so
that cost economies through linking under common
ownership appear to be small.
    The second issue is probably more important
and is concerned with the rôle played by
transactions costs. These are the costs incurred
in the search for suppliers or customers and the
costs of carrying out negotiations. These costs
can be avoided if fishing and processing are
integrated because the processing plants can use
the fish from the fishing boats and do not have to
incur costs of searching the market for supplies.
    It is argued that it was the need for
continuous supplies to keep canneries operating
throughout the season that led to backward

171

integration by processors into the ownership of boats. If the canneries did not operate continuously, costs per unit of output started to rise rapidly and it was therefore more important to secure supplies than search the market for minor price advantages.

The counter argument is that other costs might be incurred as each arm of the integrated business loses the freedom to buy or sell as it chooses through markets and to take advantage of better deals than those offered by either side of the integrated group. This may result in higher costs, particularly for the buyer.

The strength of either argument is very much influenced by prevailing conditions in the industry. When it is a buyer's market and supplies are plentiful, backward integration is less necessary for the processor. Further, under such conditions the fishing industry is likely to be in difficulties and suffering from low profits so that owning boats can be a heavy financial burden on processors, as has been discovered in British Columbia in recent years. In some cases, as a result, some divestment of fishing boats by the processors has taken place (5). By contrast, if there is expected to be a shortfall of supplies, the incentive to control boats to guarantee supplies is much greater.

Prevailing conditions affect the strength of the third argument for integration as well. This argument is again about market power since vertical integration could be a way of enabling one stage to capture the excess profit or economic rent earned by the other stage. In the case of backward integration however the problems affecting the fishing industry in recent years have much weakened this argument since there have not really been any excess profits to consider, particularly in Canada (8).

Finally, one other way for the processor to guarantee supplies, but to avoid the financial costs of owning boats is to develop informal ties with fishermen. As a result, the interrelationship between processing and fishing can be much closer than indicated by the figures of integration of ownership. A previous section noted the services provided by processors for fishermen. For instance, fishing boats have traditionally used the processors as a source of loan finance for their operations during the season. The processors offer these services to obtain the loyalty of the fishermen and guarantee supplies of fish for their processing plants. However, there are signs that in recent years the processors have found the costs of providing these services increasingly burdensome (6), particularly in British Columbia where many fishermen are unable to service their debts to processors because of their own low profits. The lack of profits in fishing makes it unattractive for the processors to rationalise the situation by buying up the boats. There seems however no easy solution to this dilemma. The processing companies are larger than the fishing companies and thus more easily able to provide finance. Processors need the fishermen to keep their own processing activities going. Without the development of specialist agencies to provide the service for the fishermen currently provided by the processors, the continued involvement of the processors in fishing seems inevitable.

6.4.2   Forward Integration

More recently, and apparently paradoxically in view of previous comments, forward integration has been taking place as fishermen and fishing co-operatives have become engaged in the processing and marketing of their own fish, by-passing the processor, whose facilities they now use on a contract basis.   In Alaska a number of native Indian fishing companies have received funds through land settlement claims, some of which have been used to invest in downstream marketing and processing. There are two reasons why the money has been spent in this way.   Firstly it has been an attempt by the fishing corporations to gain greater control over the marketing and processing of their output, recognising that efficient processing and marketing are just as important to their own success as successful fishing.   By integrating forward they have control over product handling and quality at subsequent stages and over the marketing of their products.   This may give them an advantage in marketing which the processor does not have. Secondly, the seafood business currently has attractive opportunities in markets which have expanded in the 1980s.   Forward integration has therefore been seen as a profitable investment opportunity.   This integration has been made easier by the availability of processing facilities on a contract basis.

6.4.3   The Fish Farming-Processing Relationship

Because of the different type of production system involved in farming, the relationship between fish farmers and processors is examined separately. Farmers can either sell to traders who market on their behalf (in the way that processors do for wild salmon) or they can carry out their own marketing to customers further down the distribution channel. The recent origin of the Atlantic salmon farming industry, together with the usual data problems make generalisation difficult, but some interesting issues are emerging.

Because it is possible to predict levels of supplies and to plan production with much greater certainty there is likely eventually to be less day-to-day variation in prices. Consequently the rôle of markets to adjust continuously changing supplies to demand at each stage on the route to the consumer becomes much less important. Instead there are benefits in combining the planning of farming, processing and marketing so that supplies throughout the chain can be planned in line with expected patterns of customer demand. As a result there are already signs that both farmers and processors are seeking closer relations with each other. Some use is made of contracts between farmers and trading companies to ensure a planned flow of supplies during the year and there are examples of trading companies in Norway (see Chapter 7) which are jointly owned by traders and groups of fish farmers and to which the supplies of those fish farmers are committed. There are similar examples of fish farmers extending their interests into salmon smoking and salmon processing in the Scottish farmed salmon industry.

Not all farmers in either of the two main producing countries, Norway and Scotland, are integrated forwards in this way. In Scotland for instance most of the smaller fish farmers deal primarily with wholesale markets and they have not developed these sorts of forward linkage. Indeed some of their wholesale customers find it difficult to develop efficient relationships with fish farmers because of the latter's willingness to switch customers frequently for small price gains (see also Chapter 7). Their neighbours in the Scottish trout industry, which is much longer established, have however recently extended their interests into downstream marketing and processing through the formation of a marketing co-operative. This gives them the advantage of control of a large volume of supplies, and greater control over the quality and marketing of their product. It is possible to speculate that because of these advantages, the small Scottish salmon farmers may follow suit when the disadvantages of present arrangements are more apparent to them. In this they would be following in the footsteps of the fishing and marketing companies discussed earlier which have recently been established in North America.

6.5 SALMON CURING

6.5.1   The Process And Its Importance
A considerable proportion of fresh and frozen salmon is cured before the final point of sale. By far the most common form is by smoking, carried out originally to preserve the fish but now popular for the distinctive flavours produced. Other forms of curing are by marinading and the by-products of smoking are used to make products such as pâtés and terrines. The markets for smoked and marinaded salmon have been growing very rapidly, particularly in Europe and North America.

Statistics are unfortunately virtually non-existent but it has been estimated that about 60-70% of the salmon consumed in Europe, of both Atlantic and Pacific species, is smoked and there are major smoking industries in the United Kingdom, France, West Germany, Belgium, the Netherlands, Denmark and Norway as well as in Canada and the United States. The remainder of this section concentrates on these smoking industries and on their structure and economics. It is based largely on the salmon smoking industry in Europe with which the authors are most familiar.

The sequence of processes is shown in Figure 6.2. Fillets of salmon are cleaned and then chilled and salted, or chilled in a brine solution. Small quantities of salt are used for mild cured salmon and larger quantities for hard cures with the ratio of salt to flesh in the finished fish usually 2% or more. Spices and flavours are often added at this stage to impart a distinctive flavour to the final product. After brining the product is allowed to dry to allow moisture to move to the surface producing a natural glaze in the final product. The fish are

Figure 6.2   Processing flow: smoking

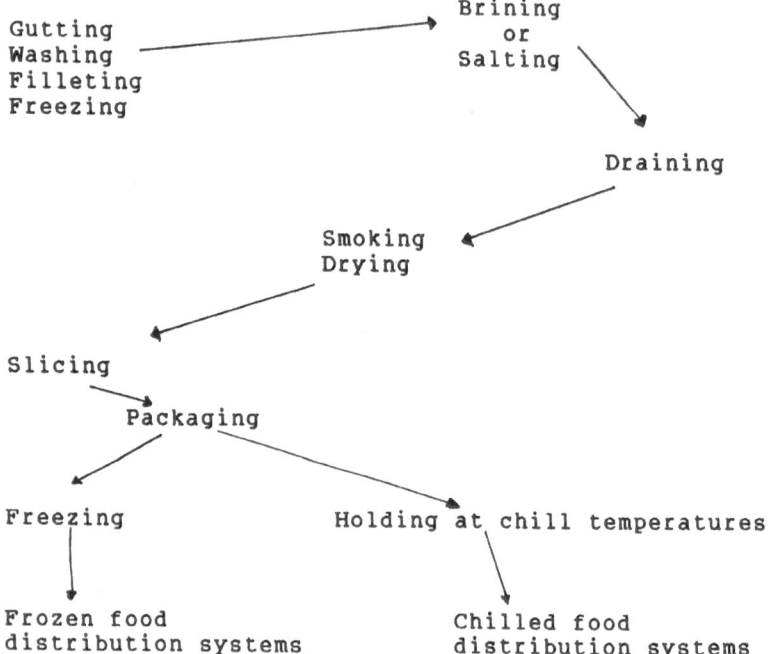

then smoked in a variety of different types of facility. These range from the traditional smoke shed, a simple building with arrangements for hanging fish from the beams and where smoke is produced by smouldering sawdust on the floor, to large commercial smoke ovens with a separate smoke generator and equipment to control temperature, humidity, smoke density and smoking time. The most common form of smoking is a cold smoke where the salmon is smoked at less than 30°C for 6 to 12 hours. Hot smoking cooks the product at temperatures of over 37°C and the taste is much stronger than that of cold smoked salmon. A typical hot smoked salmon is kippered salmon which is popular on the West coast of the United States while most salmon consumed on the Eastern Seaboard of North America and in Europe is cold smoked. The smoking process itself imparts a colour, taste and texture which varies with natural fillet quality and the length of the smoke. The weight loss during the smoking process can be 30% or more from round fish weight to finished smoked product.

Fresh smoked salmon has a shelf life similar to that of fresh fish but the use of controlled atmosphere and vacuum packaging can extend shelf life by up to 21 days. Frozen smoked salmon can be stored for many months provided that it is packed in inner moisture proof wrappings and outer protective cartons and held at temperatures of -20°C or less.

6.5.2   Markets And Marketing
The changes in technology which have extended the shelf life of smoked salmon have extended the market for smoked salmon because it can now be sold in any outlets with chilled or frozen cabinets and no longer requires the handling facilities of the specialist fish seller. The longer shelf life increases the convenience of the product to customers which also helps to extend the market.

Another change which has widened the scope of the market has been the introduction of small packs (100 to 400 grams) of pre-sliced salmon in addition to whole sides of salmon. Since both the minimum money outlay and the minimum quantity which can be purchased have fallen, the product is now more attractive to people who cannot afford a whole side or who wish to eat the product more frequently but in smaller quantities.

As the market has widened, so the

techniques of marketing have changed. Firstly,
distribution channels have in some cases become
longer since with longer storage life, the need
for quick delivery to local customers is
removed. Some local deliveries direct to retail
and catering customers remain, but it is now
increasingly common to see frozen smoked salmon
for example going through frozen food wholesalers
on the way to the consumer, many of whom are now
located some distance from the smoker. At the
same time, this does not mean that the identity of
the smoker is lost to the retailers, caterers and
consumers since the packaging, whether in frozen
or fresh form, makes it possible to give the
product a strong brand identity. This is very
noticeable in the marketing of frozen smoked
salmon which requires a rigid outer cardboard pack
for protection, providing excellent opportunities
for attractive labelling. This leads to brand
loyalty among consumers which can be further
encouraged by supporting advertising and
promotional material.

Just as channels have become longer for
some sales, so at the same time short channels to
major retailing groups have become increasingly
significant. Because the markets for smoked
salmon have been growing rapidly and because of
the improved packaging supermarket groups have
added smoked salmon to their product ranges. In
the United Kingdom for instance most of the major
supermarket groups now carry at least one line of
chilled smoked salmon, selling either under their
own label, or under that of the smoker. These
are delivered throughout the United Kingdom by the
chilled distribution network of the retailer or by
contract haulage.

6.5.3   Industry Structures

The United Kingdom pattern is taken as being
fairly typical of the general European pattern.
In the United Kingdom, where it is estimated that
around seven thousand tonnes of salmon are
produced annually, a wide variety of different
types and structures of business exist. There
are over 60 businesses smoking salmon but these
vary enormously in size with probably the usual
80/20 rule operating, i.e. around 80% of the
output comes from around 20% of firms. The
largest companies are smoking around 1000 tonnes
of salmon per year – the smallest less than 5
tonnes (see Table 6.8).

Table 6.8:   Structure of the British salmon
               smoking industry

Size of salmon smoking interests (volume of salmon purchased)	Purchases of salmon (tonnes)	Number of companies
less than 10 tonnes	150	30
11-50 tonnes	200	12
51-100 tonnes	550	7
101-300 tonnes	1200	5
more than 300 tonnes	4400	9
	6500	63

Source: Shaw and Rana, op.cit.

The corporate structures of these business are equally diverse. Some companies specialise in salmon smoking, others are involved in smoking other fish products and some are engaged in other forms of fish and food processing – no one pattern dominates, as can be seen in Table 6.9.

Table 6.9:   Interests of British salmon smokers

% of total sales volume represented by smoked salmon products	Numbers of companies	% of companies
less than 25%	17	37
26-50%	11	24
51-75%	7	15
more than 75%	11	24
	46	100

Source: Shaw and Rana, op.cit.

This variety suggests initially a lack of strong economic arguments for any particular structure, but this is somewhat misleading. By returning to the issues of markets and technology a pattern emerges from this complexity. As far as markets and costs are concerned, there appear to be advantages in being large and commanding a significant market share. Automated slicing equipment, automated production control and the use of large scale smoking equipment bring economies in costs. Labour requirements per unit of output fall with larger scale operations, as do capital costs. For instance, it is estimated that scaling up from a smoking system capable of smoking 250 kg per batch to one capable of smoking 500 kg reduces capital costs per unit of output by 25% and labour requirements by over 25%. There are similar economies in automated packing, although the extent depends on the level of labour costs which varies by country and region. A large scale of output more readily supports laboratory facilities for quality control and the cost of providing these facilities may act as a barrier to the entry of smaller firms. All these advantages are important to those customers who buy in bulk and who are very price sensitive. Consequently, those smoking businesses supplying larger retailers, caterers or wholesalers tend themselves to be large or to have grown large through the development of this side of their business.

By contrast there are likely to be two reasons why there are also many small firms. Firstly, markets are segmented and many of the smaller producers specialise, particularly in the smoking of very high quality or very distinctive products for which some customers are prepared to pay higher prices. They do not compete directly with the larger smokers supplying the volume markets. For the future, there would appear to be no reason why such firms should not survive and prosper, since demand is still growing. Secondly and by contrast, there appears to be a high turnover rate among small smoking companies and some do not appear to survive for very long. Entry into the industry is relatively easy – for instance in the United Kingdom a small smoking kiln only costs £1600 and can easily be installed in existing premises provided that they are clean and well ventilated. Because demand is growing, this seems an attractive industry to enter. Those however who neither offer the specialist product of the high quality smoker nor the low costs of the volume producers do not survive for long because it is a very competitive business.

Although there are some examples of integration between farmers and smokers, the limited extent of such integration probably reflects a difficulty in matching farming production schedules with those of smoking. Many of the smokers, particularly the larger ones, value their ability to smoke salmon of different types and would in any case not want all of the different sizes of fish produced in a typical farm. Similarly the customers and the way in which marketing is carried out are very different for smoked and fresh salmon. Thus economic cases for forward and backward integration are weak. When smokers extend their interests they have preferred to do so by expanding horizontally, i.e. into other processed seafoods to increase the range of products with which they offer customers rather than by becoming involved in farming (see Table 6.8). It is however the case (see Chapter 6) that informal relationships between farmers and smokers are becoming closer because of the increasing use of contracts between them.

Finally, turning to spatial issues, in the United Kingdom the industry is spread around the country with the largest concentration in the London area but with important industries in Scotland and the North of England. Traditionally, when smoked salmon had a very short shelf life, it was prepared by smokers located close to major markets. This explains concentrations in the London area. These businesses often operated their own distribution systems direct to retailers and caterers because of the need to get the product to customers quickly. This type of business does still exist but the development of fresh smoked products with longer shelf lives and improvements in chilled transport distribution systems have meant that the industry is no longer tied to locations close to major markets. Indeed the most rapidly expanding companies in Britain are located in rural areas, often near to their sources of supply, rather than the market.

6.6 SUMMARY, CONCLUSIONS AND FUTURE DIMENSIONS OF STRUCTURAL CHANGE IN PROCESSING

Fresh and frozen salmon are either processed by fishermen and farmers or sold to specialist processors for gilling, gutting and freezing.

Salmon for canning are mainly sold to specialist canning companies who often are also processors of fresh and frozen salmon. Generally levels of concentration in processing are modest. There is some but not universal formal integration between harvesting and processing and there is considerable informal integration of stages. Salmon smoking however is largely separate from other forms of processing.

The variety of cost structures, processing activities and markets have created a hetero-geneous collection of firms. Large and small firms, vertically and non-vertically integrated, exist side by side. Nevertheless, industry structures are often affected by the same general change factors, changes in the technology and marketing environments being the most important. Accordingly, in this final section the analysis is developed to speculate on the impact of changing environments on the future organisation of the processing industries.

A general prediction is that while opportunities for smaller firms will remain, the advantages of both size, diversification and formal or informal vertical integration will increase in all sectors. In this, salmon processing will mirror the patterns of change in food processing generally where both concentration and the extent of de facto vertical integration have been increasing (10). These changes are market rather than technology led, although technology developments have been an enabling factor. In many of the major world salmon markets changing consumer shopping habits and the development of retailing technologies are leading to increasing concentration in retailing and wholesaling. At the same time, rising real consumer incomes are stimulating a demand for higher quality food products. These two trends are likely to increase the marketing advantages of the larger processors. The size of individual orders from retailers and wholesalers is rising, so that it is an advantage for their suppliers also to be large to meet these needs. Larger firms may find it easier to meet large orders for consistent size and quality fish because of the large pool of supplies which they have available. Related to this, closer relations with suppliers may be necessary for processors to meet retailer needs for planned, orderly and quality controlled supplies. It is an advantage also to have a diversified range of seafood products since there is evidence that retail buyers are trying to reduce the number of supply points with which they deal (10).

There will always be opportunities for small new firms who see gaps in the market left by established companies. There will always be opportunities for smaller processors in specialist sectors of the market where the volumes required are small. At the same time, the changes underway in markets may require small processors to seek ways of combating size disadvantages to survive. This can be done through formal and informal groups for joint marketing and processing activities, so that the extent of co-operative ventures among smaller processors is accordingly expected to increase. In turn this may make the salmon business even more international as larger groupings will find it easier to develop international trading activities.

NOTES
1   F L Orth, J R Wilson, J A Richardson, S M Piddle, <u>Market Structure of the Alaska Seafood Processing Industry, Vol. II Finfish</u>, University of Alaska Sea Grant Report 78-14 1981.
2   R Schwindt, <u>Industrial Organisation of the Pacific Fisheries</u>, McDaniels Research Ltd, prepared for the Commission of Pacific Fisheries Policy Vancouver 1982.
3   F M Scherer, <u>Industrial Market Structure and Economic Performance</u>, Rand McNally College Publishing Company 1980.
4   J V Koch, <u>Industrial Organisation and Prices</u>, Prentice Hall 1980.
5   E Pinkerton, The dressed and the undressed, competing processor strategies in the market for raw salmon, <u>Annual Meeting of the Canadian Sociology and Anthropology Association</u>, Vancouver 1983.
6   M Shaffer, <u>An economic study of the structure of the British Columbia Salmon Industry</u> prepared for the Department of Fisheries and Oceans and the British Columbia Ministry of the Environment 1979.
7   UMA/ERA, Underwood McLellan and Associates Ltd., Edwin, Reid and Associations Ltd <u>Competitiveness and Efficiency of the British Columbia Salmon Industry</u> for the Canadian Government 1976.
8   Department of Fisheries and Oceans, <u>A New Policy for Canada's Pacific Salmon Fisheries</u> 1984
9   S A Shaw and J Rana, Salmon Smoking in the United Kingdom, <u>Institute for Retail Studies Market Report No. 1</u> 1985.
10 J A Dawson, S A Shaw, S Burt, J Rana, <u>Structural change and public policy in the European food industry</u>, European Community Forecasting and Assessment for Science and Technology Programme 1986, FAST Occasional Paper No. 103.

FURTHER READING
Clarkson Gordon, Survey of the British Columbia Fish Processing Industry, November 1983.
J Douglas MacDonald, The Public Regulation of Commercial Fisheries in Canada Case Study No 4, The Pacific Salmon Fishery Economic Council of Canada 1981: The Alaska Seafood Industry, prepared for the House Interim Committee on Foreign Investment of the Alaska Legislature.
S A Shaw, Rôle of Cooperatives in British Trout Farming, Yearbook of Cooperative Studies, Plunkett Foundation 1983
J Upton, New Technology and an Old Art, Seafood Leader, Spring 1982.
F Walker, Brining, Smoking and Drying - one way to better prices, Fish Farmer, September 1983.
F Walker, Technicalities and Hardware of Smoking, Fish Farmer, July 1983.

Chapter Seven

DISTRIBUTION CHANNELS AND THE MARKETING OF SALMON

7.1 INTRODUCTION

The distribution channels for salmon are the
business systems handling the product *en route*
from the harvesters and processors to the final
consumer. The ability of distribution channels
to handle products efficiently and meet customer
requirements as reliably and precisely as possible
makes an important contribution to the overall
success of the marketing of salmon products. If
products are handled badly and quality control is
not maintained, if delivery is at the wrong time
or of the wrong products, harvesters and
processors will be adversely affected. Prices
are affected directly by the efficiency of
operations in the channel, since (Table 7.1) over
30% of the price of the final product is derived
from the cost of activities in the channel after
the fish leave the processor. The sections below
describe and explain the structure of distribution
channels and consider how producers and processors
can market their products through the channel,
including a discussion of new product development
issues.

7.2 INSTITUTIONS

The routes taken by salmon products vary between
countries and even within the same country and the
same markets, but there are certain common
elements in all situations. One example, the
distribution of fresh and frozen Alaskan salmon in
the United States, is taken to describe the main
factors ivolved in choice of distribution
routes. Four other examples, frozen salmon

Table 7.1: Indices of value added in processing
and distribution for selected salmon
products 1983


```
Ex vessel    )
Ex farm gate ) = 100
Price per kg.
```

	US Sockeye lb can sold in US	Sockeye frozen sold in US	Norwegian farmed sold fresh in France
Ex vessel/farm round	100	100	100
Converted to processed weight	143 (70% yield)	133 (75% yield)	120
Processor selling price	430	270	143
Agent/Broker selling price	438	278	-
Wholesaler selling price	470	299	157
Retail selling price	611	389	188

Notes:
1 These are indicative of processing, handling costs and margins
only. There are fluctuations between and within years and
variations between species and fish quality.

2 The ex vessel prices and processor prices are based on Alaskan
salmo, buying prices are lower and transport costs are higher than
in locations closer to Seattle.

distribution in France and Japan, the distribution of canned salmon in the United States and export channels for Norwegian salmon, are also discussed because they are broadly illustrative of variants on the Alaskan system.

Of course, many of these distribution channels do not only handle salmon. In the case of fresh or frozen salmon they are likely to handle other types of fish as well, while canned salmon tends to move along the same channels as other canned foodstuffs. Smoked salmon is distributed with either fish or delicatessen foodstuffs. In general, the closer the stage of the channel to the final consumer, the more likely it is that salmon products will be handled in conjunction with others.

In the longest route, Alaskan fresh/frozen salmon passes through three channel members or intermediaries before it reaches the final consumer. After processing, the product is transported to regional wholesale markets, usually, but not always via the industry headquarters in Seattle.

Some processing companies use agents or brokers as intermediaries between themselves and regional wholesalers. Agents and brokers either act for the processor on a commission basis or they buy salmon which they then resell to wholesalers. Their function is to establish contact with existing and potential new wholesale customers, to make financial arrangements and to handle documentation. Processors are usually specialists in handling Pacific seafoods and in many cases salmon products will be the only ones that they handle. At this stage dealings are likely to be in bulk.

Regional wholesalers "break bulk" because they buy in larger quantities than the volumes that they deliver to their individual customers. Regional wholesalers are usually general dealers in fish who acquire many types of fish from many different sources to meet the daily or weekly requirements of their customers. Salmon is only one of a wide variety of different fish which they handle. The wholesalers carry out secondary processing such as filleting or steaking as required by their customers, although the latter may sometimes carry out such operations themselves. Their customers are retailers, caterers and salmon smokers.

Figure 7.1: Distribution channels for fresh/frozen Alaskan salmon

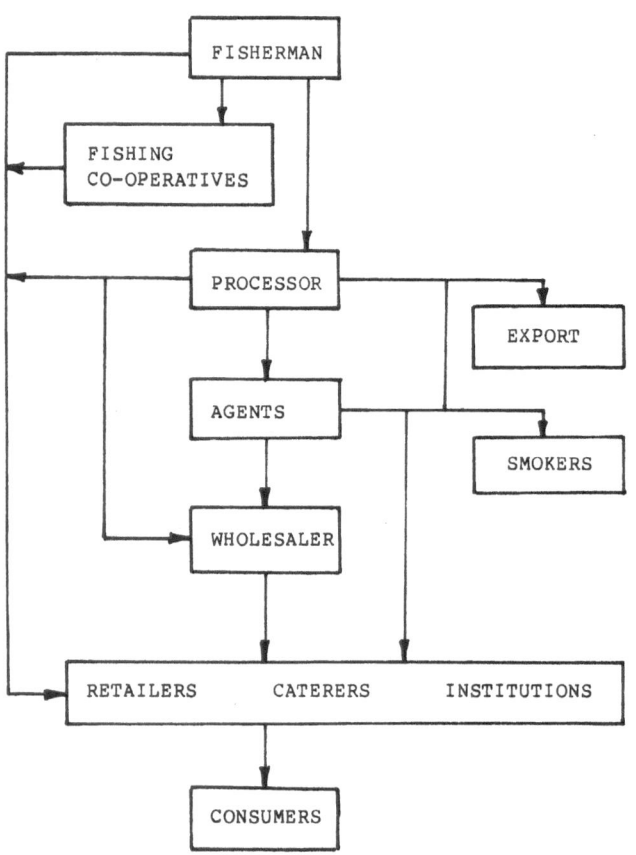

Source: industry discussions.

The pattern is somewhat similar in export markets. If, for instance, the salmon is destined for export markets in Europe, processors deal with importing companies in those countries. As before, agents and brokers are sometimes used as intermediaries between the exporter and the importer and they will be based in the market in the importing country. Agents are usually the sole representative for each exporter and do not handle the salmon of competing exporters, although they are likely to handle other fish products. The use of agents is often not considered to be ideal since the fact that the agent is handling other products may mean that the effort that they are making on behalf of the individual exporter is limited. There is however often no cost effective alternative to such arrangements. Once again (Figure 7.2) the salmon will pass to regional wholesalers in the importing country before going into the retail and catering sectors.

This long route is not always followed as Figures 7.1 and 7.2 show. Sometimes both harvesters and processors deal directly with retailers and caterers and salmon smokers. This does not only occur in the marketing of Alaskan salmon but is also common for other types of salmon as can be seen in Figures 7.3 and 7.4 which show distribution channels for Norwegian farmed salmon and for canned salmon. In fact there are usually both long and short distribution channels in most markets. We now turn to an explanation of these variations.

7.3 CHANNEL LENGTH

7.3.1. Supplies and Order Size
The number of intermediaries, i.e. the length of the distribution channel, can be explained to a considerable extent by the relationship between the size of orders and the volume of supplies handled by the harvester or processor. The greater the mismatch between the size of efficient dispatch and the size of demand by retail or catering customers, the more likely it is that wholesalers will be used as intermediaries. This explains why Alaskan salmon processors use wholesalers in regional markets. The wholesaler buys in bulk and sells to individual customers in smaller quantities. The use of the wholesaler

191

Figure 7.2:   Distribution channels for fresh/
frozen salmon imported into France

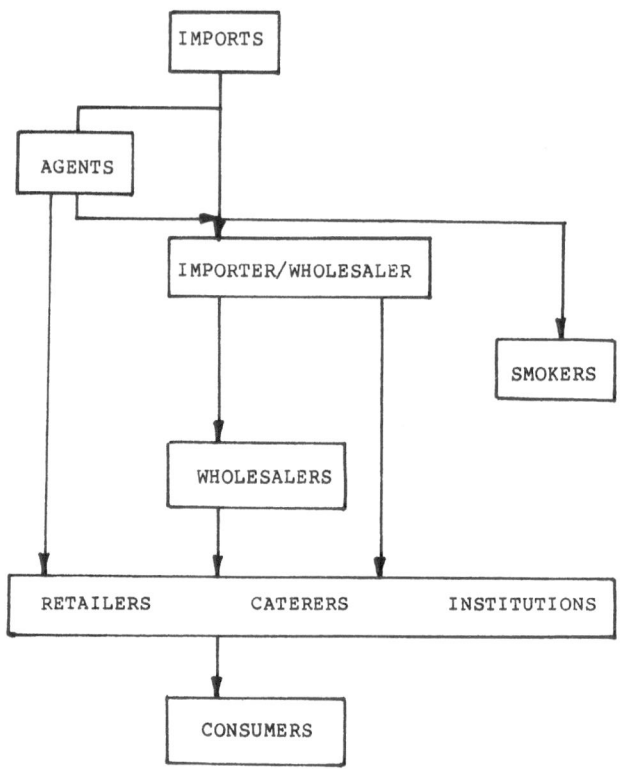

Source: industry discussions.

Figure 7.3:   Export distribution channels for
               fresh/frozen Norwegian farmed salmon

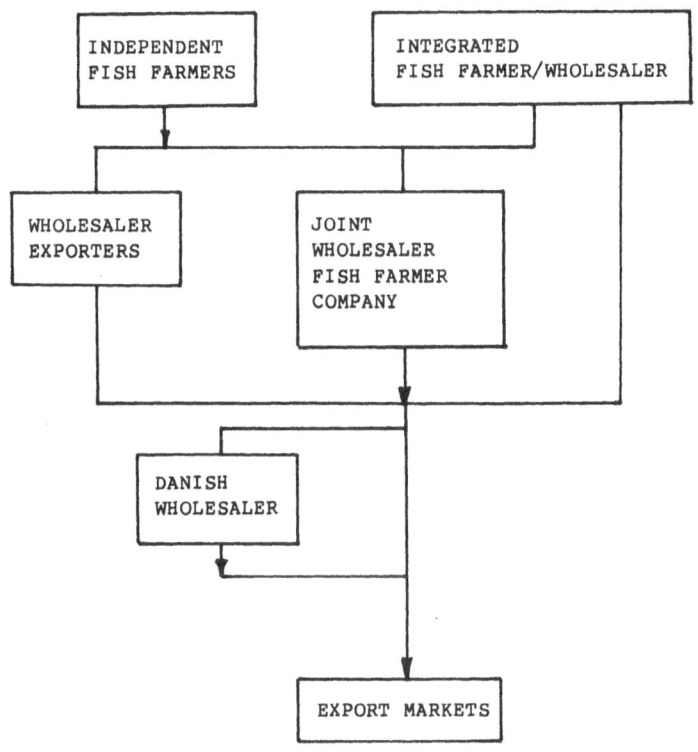

Source: Industry discussions.

Figure 7.4:   Distribution channel for Alaskan
              canned salmon in the US market

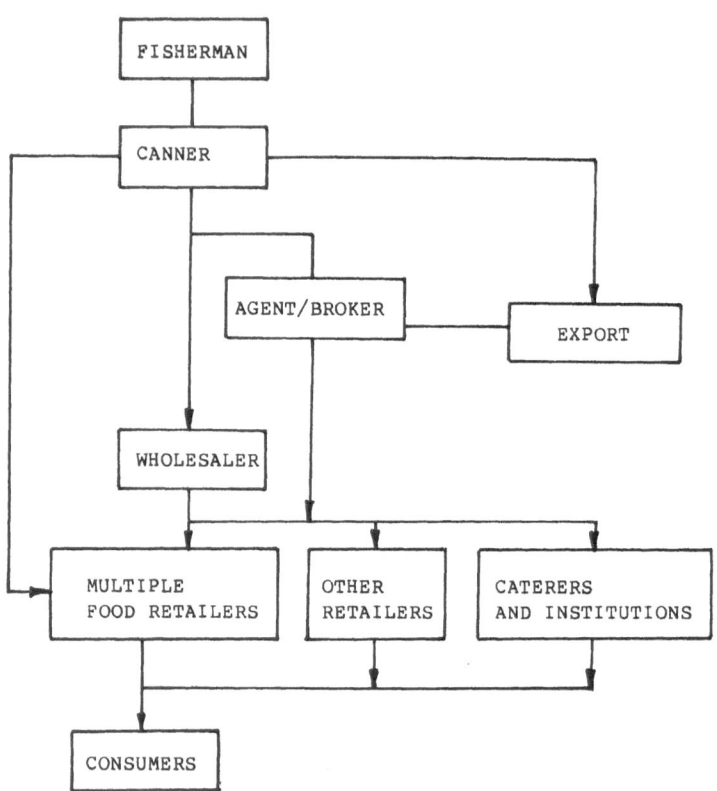

Source: industry discussions.

economises in both transport and market costs as
the wholesaler pools costs of dealing with customers across a number of different types of
fish. Where there are exceptions to this
pattern, demand by each retail and catering
customer is large enough to justify both direct
transport and direct dealings between the
processor and the retailer or caterer. This is
usually only the case for puchases by large
multiple retail and catering groups, retailers and
caterers located close to the source of supplies
or for purchases by large salmon smokers. Occasionally such direct buying can cross international boundaries. For example, some European
smokers buy salmon directly from processors in
North America and some Japanese supermarkets buy
directly from United States processors.

The size of the order is as influential as the
size of the customer. Thus, it is unusual to
find supermarket groups in the United Kingdom
buying frozen salmon directly from processors or
harvesters in North America, although the British
supermarket groups are very large companies.
Instead they obtain their requirements from
brokers or importers in the United Kingdom because
British supermarkets sell only small quantities of
frozen Pacific salmon and such small volumes do
not justify the time and trouble needed to deal
directly with exporters in other countries. On
the other hand British supermarket groups buy much
larger quantities of North American canned salmon
and these quantities are sometimes large enough to
justify direct dealing with North American
exporters.

Clearly, there will be variations between
markets and between products from different
sources. Typically, however, fresh salmon has a
longer distribution channel than frozen or canned
salmon. This is because substantial quantities
of fresh salmon reach the consumer through large
numbers of small specialist fish retailers. The
volume of their individual purchases does not
justify direct delivery and as a consequence it is
more efficient for them to buy from wholesalers
who serve their requirements on a regional
basis. Even in a country like the United Kingdom
where increasing quantities of fresh salmon and
other fresh fish are being sold through multiple
supermarket groups, these groups in many cases
still use regional wholesalers. Fresh fish have
to be separately handled from other products and

it is not cost effective for the volumes handled to set up a specialist in house distribution network to service the needs of each store.

Although a variety of patterns exists and is likely to continue to do so, the volumes travelling through shorter channels have been increasing at the expense of longer channels. This mainly reflects the increased concentration in retailing and catering which has been occurring in many countries. This has created larger retailing and catering groups, who have sought to buy more of their requirements directly from processors because they are buying in large quantities. This trend to more direct buying by large groups has been facilitated by the growth of systems of contract transport and increases in the flexibility of transport systems which have enabled them to buy products directly without having to provide transport themselves. Thus they can buy directly and avoid the wholesale margin without incurring the costs of setting up a new transport system. This reduces the minimum order size at which the time and effort involved in direct dealing becomes worthwhile.

7.3.2   Market Knowledge And The International Dimension
Dealing in international fish markets is a highly skilled activity demanding detailed specialist knowledge of markets, sources of supply and suppliers, as well as understanding of the intricacies of international currency movements. For this reason only small numbers of wholesale, retail and catering buyers consider the effort of obtaining this knowledge worthwhile and most prefer to buy through specialist importers, agents and brokers. Large salmon smokers in Europe cross international boundaries in buying but the volumes of salmon which they handle often amount to a thousand tonnes per annum and the effort is therefore justified.

These relationships are not static over time and change in response to changes in supplies, technology and markets. Improved international communications have made it easier for suppliers to deal directly with a wider variety of customers in export markets. The physical movement of fresh and chilled products in particular has become easier with the development of better equipment and improved methods of storage. This has made it easier for processors to dispatch small loads directly further down the distribution channel and to take part in export marketing themselves. It has also made it easier for processors and dealers to concentrate on marketing and customer relations and leave physical distribution to others.

7.3.3   International Integration
Few importers and exporters have crossed international trading barriers to combine both rôles by establishing a permanent presence in both the exporting and the importing country. This reflects the segmented nature of international markets. Each national market is best understood by those operating in it and the investment in sales offices in export markets would require larger volumes of sales to those markets than most exporters handle to be justified.

The major exception to this is the Japanese case. The previous chapter noted that Japanese fishing and trading companies are heavily involved through ownership ties with North America (1), (2). Japanese imports of frozen salmon are large and these acquisitions were made mainly in the 1970s as part of a general movement by Japanese companies across a wide range of products to control the resources which they imported, especially those resources, like salmon, which they expected to be in short supply (3). These companies handle not only processing operations but also onward marketing and distribution of salmon in Japan. The success of these companies has however been mixed, particularly when salmon prices have been low. As a result, it is not clear that this process of acquisition will proceed further.

7.3.4   Custom and Precedent
Finally custom and precedent play a rôle in distribution channel structures. Salmon traders are part of a general system of fish distribution and are influenced by the economy and society to which they belong. It is otherwise difficult for instance to explain the contrast between the Japanese frozen salmon distribution system (Figure 7.5) and that of France or the United States (Figures 7.1 and 7.2). The Japanese system is complex and circuitous for most products including salmon. Usually salmon passes through the hands of many intermediaries, each performing narrow and highly specialised functions. This is partly

because Japanese consumers shop very locally and very frequently and partly because members of the channel have been bound together in traditional relationships and patterns of financial indebtedness which have until recently inhibited change (4). There are some signs that this is now altering, as was illustrated earlier by the direct buying by Japanese supermarkets from United States processors, but the speed of change is slow.

7.4 RELATIONSHIPS IN MARKETING CHANNELS

The workings of distribution channels are not only explained by the patterns of legal structures but also by the nature of working relationships between buyers and sellers

A traditional view would be that each stage of the marketing channel consists of independent and autonomous buyers and sellers who switch continuously from one source of supply to another in search of the best deals. Suppliers whose prices are higher than those of competitors will lose business rapidly and be forced down to the ruling industry price level (5). Intermediaries make their profits by astute buying and selling, which require that they are well informed about prices from different sources and rigorous in switching between suppliers in search of better bargains. Thus it is by competition within the channel that consumers' needs are met, rather than through co-operation between subsequent stages.

On the whole, this view gives a reasonable approximation of the workings of salmon markets. Being competitive in price is important to the success of sellers because of the homogeneous nature of supplies. As a consequence, it is necessary for dealers to have accurate and up-to-date market information and to avoid long term commitments. This is especially true for fresh salmon distribution where most trading is still done on a day to day basis. Extensive use is still made of traditional wet fish wholesale auction markets such as Rungis in Paris and Billingsgate in London. Here retail, catering and secondary wholesalers buy from the market wholesalers at prices which change daily, in the light of day to day changes in the demand and supply situation.

In some ways however this is an over-simplification of the way in which business is

Figure 7.5:   The distribution of frozen salmon in
              Japan

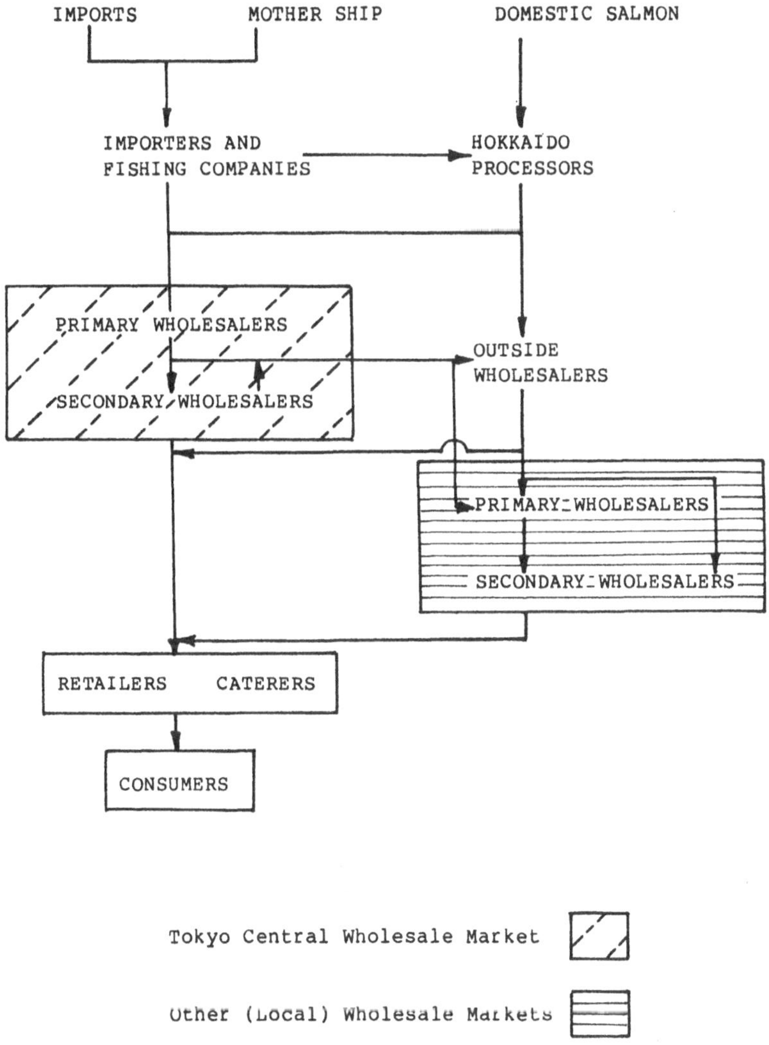

IMPORTS   MOTHER SHIP   DOMESTIC SALMON

IMPORTERS AND   HOKKAĪDO
FISHING COMPANIES   PROCESSORS

PRIMARY WHOLESALERS

SECONDARY WHOLESALERS   OUTSIDE
                                      WHOLESALERS

                                      PRIMARY WHOLESALERS

                                      SECONDARY WHOLESALERS

RETAILERS   CATERERS

CONSUMERS

Tokyo Central Wholesale Market

Other (Local) Wholesale Markets

Source: industry discussions.

done. Firstly while the emphasis on competition is probably correct, the emphasis on lack of co-operation between stages probably is not. It has never been a very accurate description of the nature of relationships, i.e. the way in which business is actually done, because the rôle of buyer-seller loyalty is ignored (6). This can be defined as the extent to which buyers tend to keep the same suppliers rather than continuously switching around in the search for better terms. Discussions with industry members suggest that this type of loyalty does indeed exist and that there is considerable co-operation between different stages in the channel. There are very good economic reasons for this. While many of the features of the product are visible to the buyer, others are not. For instance, speed and reliability of payment, delivery on time and delivery in good condition can be just as important as price. These are matters on which it can be very difficult to obtain information from a new unknown supplier or buyer. Accordingly, there are advantages in dealing with familiar buyers or suppliers in order to reduce these buying and selling risks.

This does not mean that marketing channels are uncompetitive. Buyer-seller loyalty will only last as long as each offers a satisfactory deal to the other. Within these stable relationships, each party will continue to monitor trends in the outside markets and if their partners get out of line with the rest of the industry, switching will occur. What it does mean, however, is that in practice the channels work more smoothly because of these relationships which have through time established mutually acceptable patterns of organisation for both sides. Because both sides have invested time in developig the relationship, they have accepted a need to keep their partners well informed and to work with them to help to ensure that the overall channel works as effectively as possible (7)(8).

Changes currently taking place are strengthening the need for co-operation by members of the marketing channel. One change is associated with the nature of salmon farming and has already been mentioned in Chapter 6. A barrier to co-operation in the channel in the past has been caused by the large fluctuations in production and the difficulty in predicting future supplies. Members of the channel, particularly wholesalers, have been reluctant to risk forward commitments with suppliers or customers because of the uncertainties. However, because supplies are more predictable, more use is being made of contracts to supply farmed salmon. Because the risk of price fluctuations is smaller, there is less risk in having such contracts for either side but at the same time advantages of continuity and stability are gained. Because there are fewer transactions involved with contracts, the costs of procurement are also reduced, which can be passed on to customers in the form of lower prices. This trend has been most noticeable in supplies of farmed salmon products to large retail and catering groups and to large salmon smokers (9) who prefer to do business in this way.

Smokers have particular interests in planned supplies and fixed prices. They supply their customers with price quotations many months in advance of actual sales dates, a procedure which is facilitated when they obtain their own supplies on similar terms. Fixed prices and guaranteed supplies make this much easier. With wild salmon, to achieve guaranteed supplies at known prices often meant buying during the season and storing until needed, although this tied up considerable capital and incurred interest charges. With farmed salmon these costs and risks can be avoided.

The change induced so far by farmed salmon should not however be overstated. The largest quantities are still going through the traditional wet fish auction markets where prices are fixed on a day to day basis. Nevertheless since direct buying by large customers further down the chain is becoming more important and their insistence on stable planned supplies is noticeable, the overall trend is likely to be towards increasing use of formal contracts and more buyer-seller stability.

Working relationships have also become closer because of changing patterns of consumer demand. Generally consumers today are demanding higher quality products, as are the retailers and caterers who supply them. (9) High and consistent quality is best achieved by developing strong working relationships and regular supplies from known sources. This is not making the salmon business any less competitive as anyone selling to these customers will testify. It is merely changing the way in which business is done and the way in which competition works.

7.5 MARKETING, THE MARKETING CHANNEL AND THE FINAL CONSUMER

All members of the channel from harvesters to retailers and caterers have an interest in ensuring that final consumers are aware of salmon products and view them favourably. This interest is however strongest for those most dependent on salmon, i.e. the harvesters, processors and smokers. Because of this the responsibility rests largely with them to "pull" their products through the distribution channel by heightening consumer awareness, while at the same time "pushing" the product through the channel by increasing the incentives for members of the channel to market salmon.
One difficulty however which harvesters and processors face is a lack of incentive for the individual company to promote to the final consumer because the latter cannot distinguish between salmon originating from different companies (Chapter 6). For this reason most of the marketing activities by individual producers and processors are towards their immediate customers and not towards the final consumer. A way in which this problem is overcome is through generic marketing. This is defined as the marketing of a particular species or product jointly by members of that sector of the industry, usually through their trade association and sometimes with government support. Generic marketing activities have a long history in the salmon business. In 1930 in a period of depressed demand for canned salmon, canners in British Columbia carried out a generic advertising campaign to help raise salmon sales (10) and generic work has been carried out by many different sectors of the industry since that time.
Two examples indicate the kinds of activities involved. The Alaska Salmon Marketing Institute (ASMI) was founded in 1981 with twin objectives of promoting Alaskan seafood of all types and of raising the quality of salmon exported from Alaska. It is run by a managing board of processors and fishermen and is financed by government grant and a levy on seafood companies. In 1983/4, of a total budget of US $3.9 million to cover all seafoods, 33% was spent on advertising canned salmon, 17% on fresh and frozen salmon and $120,000 on quality improvement programmes. Promotional efforts have been

directed both at members of the channel and at
final consumers, using advertising on radio and
television and in magazines, point of sale
material, displays and presentations. The work
on quality is carried on in parallel with
promotional work and involves establishing
voluntary quality control programmes among
fishermen and processors towards whom educational
programmes are also directed (11). Such
activities have helped in the successful marketing
of Alaskan salmon because they have created
interest in the distribution channel both directly
through communications with the channel and
indirectly because of the awareness of channel
members that their sales are in turn being helped
by the efforts at final consumer level.

The Norwegian Fish Farmers Sales Association
(12, 13) is a trade association to which all
Norwegian fish farmers belong. It works in
conjunction with licensed exporters in its
marketing activities, so like ASMI, it links
harvesters and processors. It has developed
detailed quality and handling specifications and
grading schemes for Norwegian salmon. It
undertakes promotional work, although given its
smaller budget (£600,000 in 1984) promotional
efforts through advertising are mainly directed at
the channel. It does however carry out a great
deal of public relations work at final consumer
level by providing material for news articles and
exhibitions and the latter have been very
successful in stimulating widespread consumer
interest. The Norwegian Association has
developed educational programmes for fish farmers
to raise their quality standards and to increase
their awareness of the need to produce year round
supplies in a planned manner. It goes further
than ASMI in suggesting minimum price levels for
salmon and it offers an extensive price
information network. The Scottish Salmon Growers
Association performs a similar function for
Scottish salmon, except that its rôle is at this
stage limited to promotional and quality control
areas.

The value of these activities is
considerable. Consumers in major markets are
faced by a multiplicity of choices of foods so
that it is important both that they are offered
products of high quality and that they are made
continuously aware of the merits of salmon in its
various forms. Generic marketing is particularly

valuable because it promotes the products of many producers at the same time and thus makes a bigger impact than that of individuals acting separately.

The success of these activities is not just a function of the amount of money spent. A further requirement is that the promotional efforts are backed by a commitment by those participating to maintain and improve product quality. Unless the product itself presents a common and consistent image to consumers, there will be a gap between the advertising image and reality which consumers will not accept. This quality need has been recognised through the educational programmes and quality control schemes mentioned above.

Quality control schemes can either be based on voluntary compliance or they can be controlled through compulsory product inspection schemes using sanctions against processors and harvesters who do not meet required standards. Whether voluntary control schemes alone are effective in establishing adequate quality standards is difficult to say at this juncture. There is always a danger when compliance is voluntary that some people will choose not to adhere to standards. For example, in Scotland, while in general quality is high and has been rising, there have been complaints from wholesalers that some farmers fail to keep to the guidelines. Farmers occasionally dispose of lesser quality fish through normal distribution channels which benefits them in the short term but which can damage both themselves and the general reputation of the industry over the longer term. For this reason the Scottish Salmon Growers Association have now established a quality control scheme which will be monitored by compulsory inspections of plant and product for those who wish to be members of the scheme. Fish of premium quality will be easily recognisable by their labelling and any industry members who fail to meet standards will be prohibited from using the quality seal of approval (14).

A final problem is created by those who do not wish to belong to the industry association or to comply with the quality guidelines. They benefit from the generic promotion carried out by the association towards the cost of which they are not themselves contributing, but at the same time if their products are not of good quality they can damage the general reputation of the product. This makes it all the more important that the

Associations establish clear and well known standards and identification methods so that buyers can
identify the members of the Association. Fortu-
nately, the heightened general awareness of the
importance of good marketing means that while the
problem exists it is possibly declining in significance.

7.6 NEW PRODUCT DEVELOPMENTS AND THE MARKETING CHANNEL

A final ingredient in the successful marketing of
salmon is good product development work. Innova-
tion through the development of new products and
new processes has always made a major contribution
to commercial success and the salmon industry is
no exception to this general rule. A small
number of examples of recent developments is
listed in Table 7.2.

The most notable feature of these developments
is that they have occurred at different stages in
the channel. For example, in the United Kingdom
the initiative to develop controlled atmosphere
sales of fresh salmon was as much due to the activities of a major retailer as to the processors.
The success of the venture involved co-ordinated
product development work by growers, processors
and retailers. Techniques of air-freighting
fresh salmon have been developed by air-freight
companies in conjunction with producers. This
re-emphasises the point that the health of the
industry as a whole depends on the successful co-
operation of all members of the chain in marketing
activities.

Given the expected increase in future supplies
(Chapter 8) a continuing process of new product
development is necessary for the future health of
the industry. Opportunities vary in different
parts of the world, but from a European perspec-
tive the scope for development still exists.
Consumer demands are changing and new food products are needed to meet new demands. In particular consumers are seeking more pre-prepared
convenience foods and seeking to buy fish from a
wider variety of food outlets, requiring further
developments in presentation and packaging.
Markets for processed and pre-prepared salmon
products in Europe, with the exception of those
for smoked salmon, have barely been developed.
The convenience markets for pre-packed portions

Table 7.2: Some recent innovations in the salmon industry

Innovations	Source	Advantages to Consumers	Examples of Impact
Salmon Farming	Fish farmers	High, consistent quality, year-round	(i) Developments of existing and new markets for fresh salmon (ii) Development of sales of Atlantic salmon to multiple retail groups
Techniques of air-freighting fresh salmon	Air freight companies Fish farmers		(i) Developments of out-of-season markets for fresh salmon in North America
Vacuum packing controlled atmosphere packaging	Packing manufac-turers Processors, Retailers	Longer, safer shelf life fresh products	Widening of market for smoked salmon to new retail outlets
Sliced smoked packs	Packing manufac-turers Processors	Smaller smoked salmon packs: lower unit price More effective use of lower quality materials: lower price	Development of sales of smoked salmon to new groups of consumers
Canned boneless salmon	Canners	Ease of preparation changed texture	Widening canned salmon market
Processed salmon products (terrines, pre-cooked dishes, etc.)	Seafood Processors, retailers	Ease of preparation wider product range	Higher rates of consumption by existing buyers Purchases by new buyers

and slices of smoked salmon can also be developed. A special advantage of these kinds of developments is that they may introduce salmon to consumers who are at present not very aware or knowledgeable about salmon at all but who are attracted by convenience and ease of use. It is even possible that this can also lead consumers, once this unfamiliarity is overcome, to more interest in the consumption of salmon in more traditional forms.

7.7 SUMMARY

Distribution channels are very diverse, with the length of the channel depending on the markets to be served, the size of customers and the volumes handled by the supplier. Long channels which involve a number of wholesale intermediaries as well as retailers and caterers are used where sources of supplies and markets are remote from each other and the size of individual supplies and orders is small. Shorter channels are more common when large retail and catering groups are involved in buying in large quantities. Because of the volatility of salmon markets, members of the channel retain the flexibility to switch sources of supply and customers as conditions change. However, because of the greater potential stability in production patterns for farmed salmon in comparison with wild salmon, as salmon farming develops more close co-ordination and co-operation will be seen between stages. The functions performed by the channel add considerable value to salmon products and their efficient execution is an essential part of successful salmon marketing. Channel members will however be encouraged in their own marketing of salmon and salmon products if they are ensured steady supplies of high standard products and promotional help from harvesters and processors to increase consumer awareness and interest.

NOTES

1   F Orth and Associates and W P Dougherty, op.
    cit., Foreign Investment in the Alaska Seafood
    Industry.
2   Quadra Economic Consultants Ltd. in associ-
    ation with McDaniels Research Ltd., Foreign
    Investment in British Columbia Fish Processing
    prepared for the British Columbia Ministry of
    the Environment 1979.
3   Commission on Pacific Fisheries Policy, op.
    cit., Turning The Tide: A New Policy for
    Canada's Pacific Fisheries.
4   M Y Yoshine, The Japanese Marketing System:
    Adaptations and Innovations, MIT Press 1971.
5   O E Williamson, Markets and Hierarchies, Free
    Press 1974.
6   R J Lent, Uncertainty, Market Disequilibrium,
    and the Firm's Decision Process: Applications
    to the Pacific Salmon Market.   Oregon State
    University Sea Grant College Program 1984.
7   J A Dawson, S A Shaw, S Burt, J Rana,
    Structural Change and Public Policy in the
    European Food Industry, European Commission,
    Forecasting and Assessment for Science and
    Technology Programme, 1986.
8   J A Dawson, Commercial Distribution in Europe,
    Croom Helm 1982.
9   D Rackham, Meeting of the Scottish Marine
    Biological   Association,   Oban   1983
    (unpublished).
10 Development Planning and Research Associates
    for   the   Office   of   Commercial   Fisheries
    Development, Department of Commerce and
    Economic Development, State of Alaska, Alaska
    Salmon Marketing Information System.
11 Alaska   Seafood   Marketing   Institute,   1984
    Annual Report.
12 Fiskeoppdretternes Salgslag, Arsberetning og
    regnskap 1983.
13 Fiskeoppdretternes Salgslag, Vedtekter 1981.
14 S A Shaw and J Rana, Markets for Scottish
    Grown Salmon, Institute for Retail Studies,
    University of Stirling Market Report No. 2,
    1985.

REFERENCES

R J Lent and R S Johnson, The marketing of Pacific
Northwest Salmon, Infofish Marketing Digest
No. 3/84.
Fish Farmer, King of Fish on film, March 1984
Fish Farmer, Salmon to get £43,000 boost, February
1983
Fish Farmer, Packed smoked salmon can push up
sales, February 1983
The Grocer, Norwegians Plan U.K. expansion, August
18, 1984
Pacific Fishing, Foreign Salmon Market – an
update, September 1984
A Wright, The Marks and Spencer approach to fish
retailing, National Federation of Fishmongers'
Conference, Scarborough, U.K., 1982.

Chapter Eight

SALMON MARKETS AND THE DEMAND FOR SALMON

8.1 INTRODUCTION

An understanding of the demand for salmon and the workings of markets is necessary for any planning of future strategies by members of the salmon business. The level of profitability at all stages is affected by the interactions of demand and supply and the level of demand will determine the extent to which it is worthwhile to increase supplies through farming and enhancement. Patterns of demand define those product forms and markets which it is most profitable to develop.

This chapter starts with a discussion of demand and supply relationships in salmon markets, followed by consideration of the main factors affecting demand, and concludes with some tentative predictions for the future.

8.2 THE WORKINGS OF SALMON MARKETS

8.2.1 Supply and Demand
Demand and the forces which shape it can best be understood by examining interactions with the forces that determine supply. Figure 8.1 shows hypothetical demand and supply schedules in different years for a salmon market which could, for example, be the ex vessel market for premium grade coho. The supply schedules S1 and S2 reflect the quantities that suppliers might offer for sale each year at different prices, where it can be seen that supplies for year 2 are more abundant at all prices than supplies for year 1. To keep the analysis as simple as possible, variations in supplies within the year because of harvesting patterns are ignored at this stage.

Salmon Markets and the Demand for Salmon

Figure 8.1: Demand and supply changes

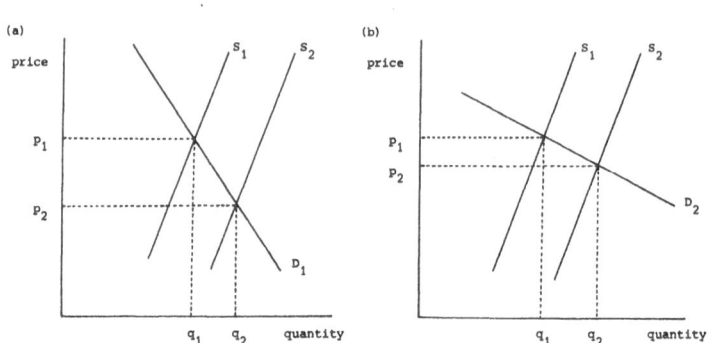

Impact on price of a shift in supply with demand curves of different elasticities

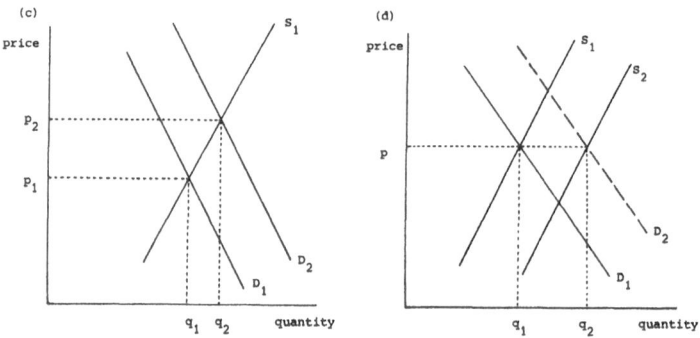

212

Both supply schedules are steep, i.e. the amount supplied is not very sensitive to changing prices. The rationale for this is that once the fishing boat is committed to the fishery or a farm is stocked with salmon there is an incentive to harvest as much as possible, whatever the level of prices.

From one year to another, by contrast, the position of the schedule is more variable. The reasons for this have been considered in more detail in Chapters 3 and 4 but are repeated here in outline. First and of particular importance in the wild fishery, there is the impact of changing environmental conditions and stock levels, leading to considerable changes in the size of runs from year to year, irrespective of price levels. Secondly, over the longer term the impact of changing technology with consequent changes in the level and structure of costs can change the profitability of operations and hence the level of fishing and farming effort. Third, if demand is increasing and prices generally are rising, through time this can lead to increased investment in fishing and farming and shift the supply curve.

The demand schedule, by contrast, represents the amount that consumers are prepared to buy at different prices. Here the schedule slopes down from left to right to reflect the fact that consumers normally buy more at lower than at higher prices. At the point where the schedules intersect, prices are determined because at this point the quantities supplied and demanded are identical. In Figure 8.1 at price P1 customers can get all the supplies they are prepared to buy at that price and this quantity is just equal to the amount that suppliers are prepared to sell. Until either the demand schedule or the supply schedule shifts this will remain the ruling market price.

8.2.2   Elasticity of Demand

If the supply schedule shifts for any of the reasons mentioned earlier, the effect on market prices depends on the response of demand which is measured by the steepness of the demand schedule. If demand is very responsive to price changes (i.e. the curve is fairly flat), a small change in prices will lead to a big change in demand. Thus, if the supply schedule shifts from S1 to S2 (Figure 8.1b) because demand is very responsive to the change, the fall in prices required to restore equilibrium prices is much less than if the converse is the case (Figure 8.1a).

The measure of the change in demand in response to changing prices is called the price elasticity of demand (see the Appendix at the end of the chapter). This is a measure of just how responsive demand is to changing prices and it is a measure of considerable significance for the industry. For example, if supply increases in the face of an unresponsive or inelastic demand curve, the industry could lose rather than gain. The fall in revenue per unit of sales as a result of the lower prices will more than offset the increased revenue from larger sales and the net effect will be a fall in the revenue received by suppliers. On the other hand, if demand is very elastic (very responsive to changing prices) there will be larger increases in sales, smaller price reductions and the revenue received by the industry will increase. If the costs of supplying the additional output are less than the increase in revenue, then profitability will also rise. The expected elasticity of the demand schedule therefore is a major element in decisions to increase supplies through investment in enhancement and farming.

8.2.3 Shifts in the Demand Schedule

Demand does not only change because of changing prices. Other causes of demand changes are represented by independent shifts in the demand schedule (e.g. from D1 to D2 in Figure 8.1c). These shifts are caused by changes in income and consumer preferences which change demand independently of any changes in prices. Thus for instance if advertising campaigns and better marketing stimulate demand, the schedule will shift to the right. Rising consumer incomes, increasing people's ability to buy high value products like salmon, will have the same effect. By contrast, if consumers decide they prefer more red meat and wish to eat less salmon, the demand schedule will shift to the left. Such shifts will also affect market equilibrium prices. As an example, in Figure 8.1c, because the demand schedule has shifted from D1 to D2 the ruling price has changed from P1 to P2. In Figure 8.1c, prices stay constant because the shift in the supply schedule has been simultaneously countered

by a shift in the demand schedule from D1 to D2.
Demand shifts which increase demand are obviously
very desirable because they increase the opportunities for the industry to increase production
without depressing prices.

8.2.4   Salmon Markets
Prices are continually changing in salmon markets
because of shifts in demand and supply schedules. The processes which have been described
above are at work in all salmon markets, both for
different types of salmon, for different salmon
products and at different levels in the chain to
the final consumer. Researchers have accordingly
sought to explain shifts in schedules and the
response of demand to changing prices so that
predictions can be made, both to define year to
year responses to changes in supplies and to
provide data to estimate the longer term viability
of enhancement and farming plans. We now turn to
the results of these studies.

8.3 THE DEMAND FOR SALMON: DEMAND ANALYSIS AND
    PRICE ELASTICITIES

8.3.1   Demand Analysis
A number of studies of salmon demand have been
carried out, but as far as possible those selected
for this chapter are disaggregated studies, i.e.
they have examined demand in single markets or for
a single product. This is because the structure
of demand varies from market to market and product
to product. There is, for instance, no reason
why the structure of demand for premium grade coho
should be the same as that for net-caught coho or
for chum. There are likely to be interrelationships between demand for salmon of different types
and these are discussed below, but there are also
differences which are obscured if more aggregated
studies are used.
   Most of the studies which are cited are concerned with demand for the Pacific salmons; largely because until recently the small volumes
available and the poor statistics have made econometric analysis difficult for Atlantic salmon.
Key factors affecting demand are shown in Table
8.1, and the most important are discussed further
in the sections below.

Table 8.1: The demand for salmon

VARIABLE	EFFECT	COMMENT
1. Own price	Higher prices reduce demand and vice versa	Key importance of value of coefficient (elasticity) As a relatively high priced fish, effect of price on demand significant particularly for higher quantity species. Elasticity likely to vary between species and product forms.
2. Price of salmon of other species and product forms	Increasing relative price of substitutes increases demand	Species not perfect substitutes for each other but substitution of adjacent species in price hierarchy. Very limited relationship between demand for salmon in different product forms.
3. Prices of non salmon substitutes	Increasing relative price of substitutes reduces demand	Canned tuna has been identified as substitute for canned salmon. For fresh/frozen and smoked salmon many other fish and meat products collectively act as substitutes but are individually difficult to identify.
4. Real Income	Increasing real incomes increase demand	Identified as major influence on demand for most product forms and species.
5. Exchange Rates	Relative appreciation exporter currency depresses demand by importer	Falling yen to US dollar and European currencies to dollar relationships depressed US prices 1980s. Affects competitive relationships as well as overall demand.

Table 8.1 continued

2

VARIABLE	EFFECT	COMMENT
6. Tariff Barriers	Increases costs and depresses import demand	Tariffs on non-processed salmon generally low. Higher tariffs on more highly processed products. Differential impacts eg Scottish smoked salmon enters EEC free. Other smoked salmon faces 15% tariff.
7. Non-Tariff barriers to trade	Reduce Import demand	Difficult to identify but include special hygiene requirements and discriminatory labelling often used to protect home industry from imports.
8. Changes in preferences	Increase or reduce demand	Changes normally slow. Most noteable recent changes: growing preference for fresh/frozen salmon over canned salmon, preference for fresh over frozen salmon. In Europe increasing popularity of fish products generally. Can affect species demand as well as product form. Difficult to predict and measure.
9. Action by suppliers:	Stimulates demand	Improved product quality and consistency — These characteristics important in developing markets for high priced farmed Atlantic salmon
		New product developments — Examples: small vacuum packs of smoked salmon.
		Positive marketing efforts including promotion — Affect not only overall demand but differential impact on products promoted. Successful penetration of European markets by Norwegian salmon in part due to extensive promotion collectively and by individual exporters.

8.3.2   Price Elasticities and the Rôle of Substitutes
8.3.2.1   Competition between species.

The value of the price elasticity of demand for a species is affected mainly by substitution relationships with other salmon as well as with other fish and food products. Substitution relationships are complex because they occur at so many different levels, from substitution between species, between different forms of processing to substitution between different qualities of fish. Substitution effects between different species are considered first.

The different species of salmon are far from being perfect substitutes for each other. If they were, processes of competition would lead to identical prices for different species but in fact as Figure 8.2 and Table 8.2 show, a hierarchy exists.

Atlantic salmon, chinook and sockeye normally sell at higher prices than coho and in turn coho sells at prices above pinks and chum. These differences reflect different perceptions of quality by consumers as well as the taste preferences of different markets for different species. For instance, a European smoker wishing to produce a high quality smoked salmon product might regard chinook, Atlantic salmon and coho as substitutes and make choices partly on the basis of relative prices, but he would be unlikely to consider pinks and chums for that market sector. In the same way, although smokers do substitute net caught fish for troll caught fish when prices change, they only do this for their more price conscious markets. Market barriers also reduce substitution effects. Devoretz (1) in a study of the demand for Canadian salmon examined the price relationships between net caught sockeye used for canning and troll sockeye going for freezing and did not find any substitution relationship. He attributed this to the ban on export of net sockeye from Canada which has the result that net sockeye does not compete with troll sockeye internationally.

While, however, substitution relationships are not perfect, they do exist and they can be quite strong, particularly between adjacent species in the price hierarchy. Movements in the demand for one species therefore cannot be understood without also understanding what is happening in the markets for other species. Muse (2) for instance

Figure 8.2:   Indexes of ex vessels prices for
              salmon in the United States

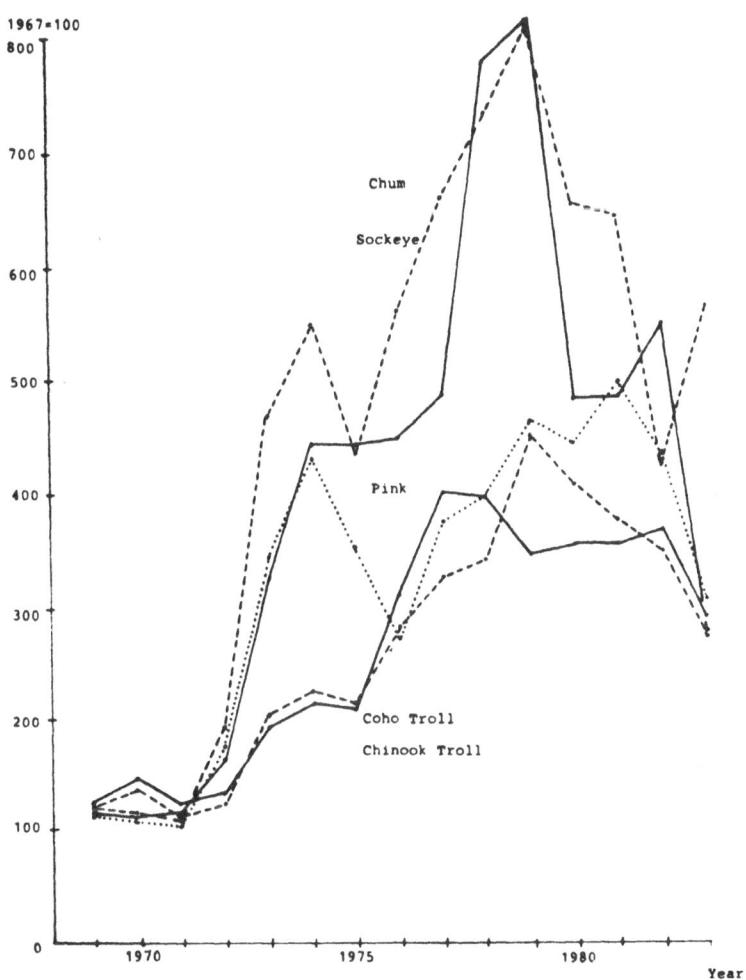

Source: Dept. of Commerce, U.S.

Table 8.2: Fresh/frozen salmon prices

	£ per kg January 1983			
Species and Weights (1b)	FOB Seattle	CIF New York	CIF Tokyo	CIF Le Havre/Boulogne
Chinook Red Troll				
4 - 7	2.92	3.12	3.12	3.70
7 - 11	3.57	3.77	3.17	4.55 - 4.74
11 - 18	4.16	4.42	-	4.81 - 5.84
18 +	4.58	4.80	-	5.78 - 6.04
Chinook White Troll				
4 - 7	1.49	1.69	-	-
7 - 11	2.14	2.53	-	-
11 - 18	2.73	2.99	-	-
18 +	3.15	3.31	-	-
Coho Troll				
2 - 4	1.94	2.08	2.08	2.12 - 2.21
4 - 6	2.08	2.27	2.27	2.26 - 2.76
6 - 9	2.22	2.40	2.40	2.40 - 2.92
Coho Gillnet				
2 - 4	1.82	2.01	2.01	2.05 - 2.11
4 - 6	1.98	2.21	2.18	2.14 - 2.34
6 - 9	2.14	2.40	2.34	2.27 - 2.53
Chum Netted Silverbright				
4 - 6	1.81	2.01	2.01	2.05 - 2.14
6 - 9	1.95	2.14	2.14	2.19 - 2.21
9 - 12	2.09	2.27	2.27	2.33 - 3.06
Chum semi-bright				
4 - 6	1.61	1.82	1.82	1.95
6 - 9	1.75	1.95	1.91	2.14
9 - 12	1.88	2.08	2.01	2.23

Table 8.2 continued.

Species and Weights (lb)	FOB Seattle	CIF New York	CIF Tokyo	CIF Le Havre/Boulogne
Pinks troll				
1 - 2	1.56	1.82	-	2.08
3 - 4	1.82	2.01	-	2.47
Pinks netted				
1 - 2	1.43	1.62	1.56	1.95
3 - 4	1.72	1.92	1.75	2.36
			1.92	
Norwegian farmed				
2 - 4	-	-	-	3.08 - 3.41
4 - 6	-	5.0	6.50	3.41 - 3.75
6 - 9	-	-	-	3.75 - 4.11
Scottish farmed				
2 - 4	-	-	-	3.41 - 3.51
4 - 6	-	-	-	3.81 - 4.05
6 - 9	-	-	-	4.50 - 4.60

Source: reference (1)

Notes: These prices were quoted during January 1983. It was a period of considerable instability as processors unloaded high stocks. Some were spot prices, others based on earlier orders. Thus they are broad indications of prices to illustrate the nature of price structures only.

found that ex vessel prices of Alaskan salmon were influenced by Japanese landings of chum: as chum landings in Japan rise, so the demand for Alaskan salmon in Japan falls, as does the price of Alaskan salmon. For instance, consumers sub-stitute canned pinks, sockeye and chum for each other as price differentials change. This is shown in Table 8.3, which is taken from a study of the export demand for Canadian salmon, again from Devoretz (3). Own price elasticities (the res-ponse of demand to a change in the own price of the good in question) were high as were cross price elasticities (the response of demand to a change in the price of other types of canned salmon).

Table 8.3: Elasticities of demand for canned salmon

	Own price elasticity of demand	Cross price Elasticity of demand
Canned sockeye	-13.8	9.2
Canned pinks	-7.3	1.4
Canned chum	-12.9	10.4

Source: Devoretz (3)

When a change in the price of one takes place relative to the price of others, consumers will switch their purchases, although as can be seen the response for each species is different, with consumers being apparently more willing to switch from sockeye and chum than from pinks.

It is not possible to present data for all species and for all markets because adequate data is not available. The examples above, however, do illustrate the need to regard each species as part of the larger market.

8.3.2.2   Competition between types of salmon.
Competition between types of salmon is probably less strong than competition between species. Devoretz (1) found, for instance, in a study of the British market that there is no evidence of any association between the market for frozen salmon and the market for canned salmon: changes in the price of one did not appear to affect the

demand for the other. This is possibly because in the British market, fresh and canned salmon are sold to different consumers and consumed in different types of meals (4).

As another example, competition between fresh Atlantic salmon and frozen Pacific salmon at retail level in Europe is indirect since the two products tend often to be sold in different types of retail outlets. Nevertheless, some substitution must take place and one example is of the trend in North America and Europe for consumers to prefer fresh salmon at the expense of frozen salmon and canned salmon, as part of a general movement towards the consumption of more fresh foodstuffs. In consequence, although not the most significant effects on demand, such influences cannot be ignored.

8.3.2.3 Competition with other foodstuffs.
Finally salmon competes with other foodstuffs, although relationships are difficult to quantify because of the wide variety of substitutions which are possible. Substitutes for canned salmon have again received the most attention, mainly because statistical data is available for use in econometric analyses. For canned salmon most researchers have found a significant relationship with canned tuna (5). For fresh and frozen salmon it has been more difficult to find relationships largely because of the poor database, but a wide variety of loose substitution relationships is likely to exist. For instance, in Europe smoked salmon is usually served as an appetiser and competes with other appetisers as diverse as fresh fruit and soups as well as with other types of smoked fish. When served as a main meal salmon competes with a range of red and white meats as well as with other types of fish.

Of course, substitutes will vary in different markets, as different cultural and social patterns dictate different responses to changing prices. The contrast between the consumption of fresh salmon in Europe and in Japan is an example here. In Europe fresh salmon has been regarded as a luxury product, selling at a relatively high price. In Japan, chum salmon is seen as a much more staple and less exciting item in the diet. As a consequence, although reliable estimates do not exist, it has been suggested that demand is much more price responsive in Japan than in Europe. Interestingly, as the quantities of fresh Atlantic salmon available have increased in Europe and the types of people buying salmon have also widened, price elasticities of demand in Europe are also thought to be rising (4).

8.3.3   The Effect of Income Levels on Demand

There are many other determinants of demand but probably the most frequently examined has been the effect of changes in consumer incomes and spending power.   Generally speaking, rising real incomes increase demand and falling real incomes depress it, a result confirmed by most of the studies made of different species and markets (5).   Like price elasticities, however, considerable variation can be expected in the actual degree of responsiveness of demand to changing incomes between species and between markets.

The rôle of income changes as well as the variation in effects across markets is illustrated in Table 8.4.   The measure used for the respons-iveness of demand to changing incomes is known as the income elasticity of demand and it is exactly analogous to price elasticities (see Appendix at end of chapter).   In Table 8.4 demand changes have been compared with changes in gross domestic incomes in each country.   It can be seen that for all products and all markets except West Germany, rising demand increases the demand for Canadian salmon and vice versa.   (The anomalous result in West Germany may possibly reflect the fact that rising incomes there have stimulated demand for Atlantic rather than Pacific salmon.)   Demand is more income elastic for fresh and frozen than for canned salmon.   Perhaps the most significant finding, however, is the size of the differences in the values of the coefficient between species and between markets.   For instance, the income elasticity of demand for all frozen species is higher in the United Kingdom than in other European markets but even within the United Kingdom variations between species and products are considerable.   This supports the earlier contention that it is necessary to separate salmon markets when analysing demand.

As well as reflecting differences in national preferences, the effect of income changes on demand will depend on existing levels of consumption of salmon, i.e. the degree of market maturity in each country.   Increases in incomes in countries where consumption is currently low are likely to lead to larger increases in demand

Table 8.4   The export demand for Canadian salmon

Price dependent demand equations: linear form

Estimated income elasticities of demand[1]:

	France	U.K.	Italy	Sweden	West Germany
(a) Fresh/Frozen Canadian Salmon					
aggregate all species	31	3.9	3.3	1.0	-0.52
	(2.8)	(5.6)	(1.8)	-	(-.51)
Chinook	1.4	3.8	3.5	0.41	0.64
	(5.4)[2]	(9.2)	(2.0)	(1.4)	(2.7)
Coho	1.3	4.0	3.3	n.a.	-0.31
	(5.3)	(5.6)	(1.8)		-0.59
Chum	1.1	2.3	1.01	1.00	-0.006
	(5.8)	(5.6)	(.5)	(8.5)	(-.02)
(b) Canned Salmon					
aggregate all species	0.58	0.23	1.7		
	(4.1)	(1.2)	(1.7)		
Chum	0.64	1.5	0.20		
	(7.4)	(2.6)	(0.21)		
Pink	0.95	1.75	1.13		
	(2.8)	(4.9)	(2.4)		
Sockeye	n.a.	2.5	n.a.		
		(4.4)			

1   This extracts results for one variable in the model only.   For full details see DeVoretz (1985).

2   T Statistics in parenthesis.   Significance levels omitted for simplicity.

than would be the case in countries where consumption is already at a higher level and the
market is saturated. Some support for this hypo-
thesis comes from the work of Devoretz already
cited in Table 8.4. For instance, one possible
reason for the higher income elasticities of
demand for Canadian frozen salmon in the United
Kingdom than in France is because the existing
levels of consumption of salmon are much higher in
France than in the United Kingdom.

8.3.4 Exchange Rates, Tariffs and Non-tariff Barriers

Exchange rates have a major influence on the
demand for commodities such as salmon which are
traded internationally (see Table 8.5). Depreci-
ations in an importer's currency relative to the
exporter's currency have both an absolute and a
relative effect on demand. First, looking at the
absolute effect, the depreciation in the
importer's currency is roughly equivalent to a
fall in real incomes in that country since the

Table 8.5:   Hypothetical effect of changes in
              currency values

British £ v US dollar

Effect of changing values assuming no change in
market prices in US dollars.   Price per Kg in UK
wholesale market.

	£1=US$2.30	£1=US$1.30	£1=US$
Atlantic salmon	4.6	4.6	4.6
Troll coho	2.6	4.6	5.9
Troll chinook	4.6	8.2	10.7

international spending power of that country is
reduced and thus demand falls (the demand curve
shifts to the left). The effect on prices in
turn depends on the importance of that export
market as a source of demand. For instance, the
appreciation of North American relative to
European currencies, because Europe is a major
market for canned and frozen salmon depresses

North American salmon prices. The converse is true if European currencies are appreciating relative to the dollar.

Turning to the relative effects of currency changes, movements in exchange rates affect the prices of different species when exchange rates of different countries do not move together. A good example of this is the relationship between the prices of the Pacific salmon and the price of Atlantic salmon. In general there is a tendency for European currencies to move together against North American currencies. Thus when European currencies depreciate or appreciate relative to the dollar, prices change and in recent years these fluctuations have often been large. Between 1982 and early 1985 the value of sterling in terms of the dollar depreciated 37%. In 1982 the price of Atlantic salmon was roughly equivalent in Europe to the price of medium sized troll chinook and was more expensive than most types of coho. As a consequence of the change in currency values, by early 1985 there were periods when the price of troll-caught chums had risen to equal that of Atlantic salmon. Since, as has been argued earlier, these salmon are substitutes in European markets, this has considerable effects on demand. Because such fluctuations are difficult to predict, this can also cause customers to seek supplies in currencies moving with their own as a way of avoiding risk. For example, in discussions with salmon smokers in Europe, a number gave this as a reason for buying increasing quantities of Atlantic salmon from Europe as a substitute for Pacific salmon.

Patterns of international demand are also affected by impediments to trade through tariffs, quotas and other similar barriers. The impact depends on the policies of importing countries but in general, tariffs inhibit trade in processed salmon products to a greater extent than trade in fresh and frozen products, particularly when the products concerned are high value ones such as smoked salmon. There are other barriers which are less obvious than this, but nevertheless influential. The structure of grants and low interest loans to salmon producers may affect relative competitiveness and helps those who receive the greatest amounts of aid. So far raw salmon and processed salmon have escaped the price support systems which regulate international trade in many sections of agriculture but informal

support through such grants and loans certainly
exists.

8.3.5   Tastes,   Preferences   and   the   Marketing Response

Finally, behind the response to changes in prices
and incomes lie questions of the degree of liking
that people have for salmon and salmon products.
Again, and not surprisingly given the large
cultural differences between the main salmon
markets, the picture is one of great variety. As
Chapter 2 has already shown, the international
differences in consumption per head and by species
are large.

These preferences can however be influenced by
the actions of the suppliers themselves. The
previous chapter has suggested a number of ways in
which such activities can be organised. Work on
quality control and new product development is
particularly important since it contributes
directly to making the products more attractive
for consumers. It is also an area where in North
America and in Europe there has been a much
greater awareness in recent years of the need to
develop industry guidelines to improve handling
standards. For canned salmon, North American
processors have had rigorous quality standards and
product inspection schemes for many years which
have recently been further strengthened. For
fresh and frozen salmon, the lead has been taken
by the Norwegian Salmon Farmers Sales Association
in the marketing of Norwegian farmed salmon, but
this lead is being followed by similar initiatives
in Scotland and in North America.

Promotion is the other major way in which producers can themselves help to stimulate demand for
their products. Such activities have a long
history in salmon canning but these activities
have now been extended to the marketing of other
salmon products, in particular by increased
advertising of fresh and frozen salmon by
associations of producers. These activities were
discussed in Chapter 7 and if the world markets
are to absorb increasing quantities of salmon
without unduly depressing prices, they are likely
to become increasingly important.

8.4 SALMON MARKETS IN THE 1990s

8.4.1 Prediction

When future prices and market conditions are discussed, the question which is most frequently raised is of the impact of the additional supplies which are expected from enhancement and salmon farming. Table 8.6 is based on estimates of known plans for enhancement and farming to 1990 and indicates the substantial nature of these expected increases, with further increases expected through the 1990s. Supplies of Atlantic salmon are expected to double by 1990 at the same time as supplies of farmed Pacific salmon are also expected to increase and wild catches are maintained through enhancement programmes.

Whether these expected increases can be absorbed without substantial declines in salmon prices depends on the demand responses considered earlier in this chapter. Firstly, it depends on the potential for non-price led growth and secondly it depends on how responsive demand is to falls in price, i.e. the price elasticity of demand.

Accurate prediction is difficult. Firstly, this is because of the difficulty already mentioned in trying to generalise across many markets, each facing different conditions. Secondly, it is difficult to predict movements in national incomes, exchange rates and preferences, all of which are key determinants of the potential for non-price led demand growth. For these reasons, rather than attempting quantitative estimates of future demand, subsequent analysis concentrates as a second best approach on identifying the key issues which will affect outcomes. The general situation is considered first, followed by a discussion of the relationships between demand and supply for different species.

In general, there appear to be reasonable grounds for optimism on some fronts. The evidence from previous sections suggests that the demand for salmon is, in general, responsive to rising incomes. If, therefore, incomes in major markets continue to rise, the demand for salmon will continue to increase. In Europe and in many parts of North America, present rates of consumption are low, particularly compared with those in Japan. The challenge in these markets is not only to stimulate demand among the most affluent for fresh salmon but also to persuade other groups that salmon is a luxury but one that is affordable

Table 8.6: World salmon supplies to 1990: Projections

tonnes

	1983	1984	1985	1986 (estimated)	1990 (forecast)
Atlantic salmon: wild total	7518	7000	7000	7000	7000
Atlantic salmon: farmed total	17,795	26,748	37,548	53,650	109,000
of which from					
Norway	14,956	22,196	28,655	40,000	80,000
Scotland	2,539	3,912	6,921	10,000	21,000
Ireland	300	340	722	1,400	2,000
Others		300	1,250	2,250	6,000
Atlantic salmon total	25,313	33,748	44,548	60,650	116,000
Pacific salmon: wild total	663,209	696,700	640,000*	650,000*	670,000
Pacific salmon: farmed total	3,550	5,600	7,200	10,600	18,000
TOTAL ALL SALMON	692,072	736,048	691,748	721,250	804,000

* estimates

Sources: Department of Agriculture and Fisheries for Scotland
        FES
        NMFS
        FAO Fisheries Yearbook.

by many. Salmon is highly regarded by consumers
and rising real incomes will enable more people to
afford and enjoy salmon. Even in Japan where per
capita consumption is so high, continuing income
and population growth are expected to stimulate
some further increases in demand. New prepared
salmon products in various forms will also help to
stimulate demand. There are opportunities to
develop markets in countries such as those in the
Middle East, the Far East and possibly Australia
if present import restrictions are lifted. The
more these opportunities are developed, the less
pressure will be exerted on prices. If downward
pressure on prices does occur, the evidence
presented earlier suggests that, given salmon's
present high price in many markets, demand would
be very responsive to price reductions, thus
limiting the extent of those price reductions
which might occur.

The picture presented therefore is reasonably
optimistic and it is of course on this basis that
the plans to increase production which are shown
in Table 8.6 are based. The major problem
however is that while we know the direction in
which demand will change, no one knows the extent
of any growth of demand that will occur and the
points at which market saturation will be reached
in the different markets. We also do not know
the degrees of responsiveness of demand to price
changes since these wil change as consumption
levels rise. If market saturation is reached at
lower rates of consumption than expected, the
price declines necessary to absorb increased
supplies could be considerable. The danger in
the present scale of developments therefore is
that the ability of the market to absorb these
supplies has been or will be over-estimated.
Physical resources available to develop farming
and enhancement are considerable - the limit of
the development of the industry is likely to be a
demand rather than a resource one. It is however
a characteristic of demand that saturation points
can sometimes be reached quite suddenly (4),
leading to rapid price declines and the emergence
of excess capacity. Further, problems of
over-capacity in the industry, if or when they
occur are likely to be especially difficult
because of the long time lag before supply
adjustments can be made. Once fish are stocked
in farms, virtually regardless of price, the
optimal commercial strategy is to rear them to

marketable size which can take up to two years. Even then because the investment in facilities is a sunk cost, there will be an incentive to continue farming in the hope of recovering some of the capital costs and in the hope that conditions will improve. Similarly, it takes at least seven or eight years to achieve the full returns to enhancement projects and once again, once the major investments have been made in facilities, the tendency will be to hang on hoping for an improvement in conditions. Thus it can take many years for excess capacity to disappear and during this time the industry would be likely to experience falling prices and profitability.

Whether this scenario faces the industry over the next decade will depend on the industry itself. Present increases planned in both farming and enhancement are considerable and some overshooting seems inevitable. It is obviously in consequence important that suppliers plan cautiously in the light of these uncertainties and that those supply increases which do occur are accompanied by strenuous efforts to create demand through promotional and new product development programmes.

8.4.2   Species Competition

Price relationships will also be affected by increases in supplies, although it seeems likely that chinook and Atlantic salmon will remain, in general, the most highly priced species. Chinook has not had a significant history of culture, and is available in the wild in only limited quantities. The price position of Atlantic salmon depends on the costs of farming since wild supplies are limited. For salmon farming to be viable implies that prices have to cover the costs of feeding and management which are typically higher than in wild fisheries. Without radical, and currently unexpected, further breakthroughs in technology, farmed salmon cannot be offered at prices comparable to those for wild fish caught through purse and seine net fishing. If prices for farmed Atlantic salmon fall below this level, over the longer term supplies of this salmon will fall back. For coho, the position is more problematic. Farmed coho, like Atlantic salmon, must command relatively high prices to justify the farming exercise but more progress has been made with the enhancement of coho stocks than with some other species and the outcome in terms of price position will depend to some extent on the balance of these two forces. Net caught pinks and chum are likely to continue to be available in large quantities through fishing and are therefore likely to remain at the bottom of the price hierarchy. The position of troll caught pinks however differs from this, since the high quality of the product gives them a better position in the market than net caught fish.

Some lesser changes in relationships are however likely to occur. We have already seen the price of Atlantic salmon in Europe fall substantially because of the increasing availability of farmed salmon. This process is likely to continue as supplies increase further. In Europe this is likely to mean that European Atlantic salmon will become increasingly competitive with imported Pacific salmon and reductions in exports of Pacific salmon to Europe can be expected.

Unfortunately, one major variable affecting species relationships – the pattern of international exchange rates – is difficult to predict and is in any case outside the control or influence of the international salmon business. Exchange rate movements affect competition between the Atlantic and the Pacific salmons, influencing day to day movements in trade and long term locational patterns of investment in the industry. Whether, for instance, over the longer term it will be cheaper to supply the United States market with Atlantic salmon farmed in the United States or Canada or from farms in Europe will depend on exchange rates. Whether it is cheaper to supply the United States market with farmed coho from the Western seaboard or with Atlantic salmon from Europe will also depend on exchange rates. Experiences in the 1970s and 1980s have been of large and unpredictable variations in currency values. It is perhaps ironic therefore that one of the most important variables affecting competition and price relationships between species is not only one which is outside the control of the industry but one which it is so difficult to predict.

8.5 SUMMARY

Key factors affecting the demand for salmon in general and by species are the effects of changing preferences, incomes and prices. It is however difficult to analyse these forces in a general way because of the wide variations in patterns of demand between species, product forms and markets. At the same time, there are also strong interrelationships between the demand for different species, so that as a consequence, markets for individual species and products cannot be fully understood without understanding these interrelationships. Because growth in incomes is expected and demand is not yet thought to be saturated in the major markets, future prospects for growth of demand are good. The main threat, however, to the industry's ability to exploit this growth profitably is the danger of excessive or too rapid increases in supplies from farming and enhancement.

Figure 8.3: Selected international exchange
rates, 1970-84

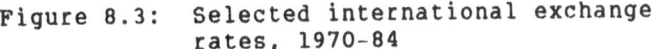

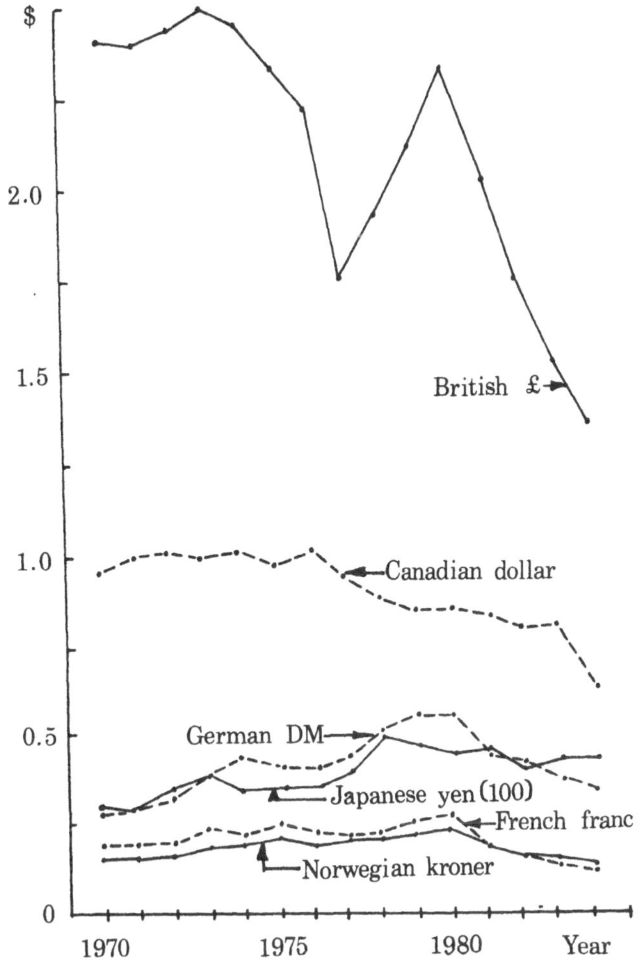

Source: derived from International Financial
Statistics (various), International Monetary
Fund.

APPENDIX
1. PRICE ELASTICITY OF DEMAND
Price elasticity of demand is used as an indicator
of how total revenue (price times quantity sold)
changes when a fall in price levels leads to an
increase in quantity demanded. It is a
percentage based measure and independent of the
units used to measure price and quantity so that
it is a measure that can be compared across
products. The definition is:

$$\text{Elasticity coefficient} = \frac{\text{\% change in quantity demand}}{\text{\% change in price}}$$

Elasticity values are divided into one of the
following three categories:

a ELASTIC DEMAND
This is when the coefficient has a value
greater than one. This means that when
prices fall the total revenue received by
suppliers increases, i.e. the % change in
demand is greater than the % change in
price. Note however that this does not in
itself mean that suppliers will benefit from
falling prices. This will depend also on the
structure of costs and in practice demand
elasticities normally have to be substantially
greater than one for the revenue increase to
exceed additional costs associated with
supplying larger outputs, i.e. for
profitability to rise.

b UNITARY ELASTICITY
The value of the coefficient is equal to
one. The % changes in price and quantity
demanded are equal so that total revenue is
unchanged.

c INELASTIC DEMAND
The value of the coefficient is less than
one. The % change in quantity demanded is
less than the % change in price which induced
it, so that if prices fall, revenue received
also falls.

2 INCOME ELASTICITY OF DEMAND
This is a percentage based measure of the
responsiveness of demand to changes in incomes.
It is defined as:

Income elasticity = % change in quantity demanded
of demand   % change in incomes.

The higher the value of the coefficient, the
greater the increase in demand brought about by a
change in incomes.

NOTES

1   D Devoretz, Export Demand for Canadian
    Salmon: EEC and Japan, unpublished, Simon
    Fraser University, 1985.
2   B Muse, Ex-vessel Price Models for Alaska
    Salmon, Commercial Fisheries Entry Commission,
    Juneau, Alaska, August 1984, Discussion Draft
    84-6.
3   D Devoretz, An Econometric Demand Model for
    Canadian Salmon, Canadian Journal of
    Agricultural Economics, Volume 30, No. 1,
    March 1982.
4   S A Shaw and J Rana, Markets for Scottish
    Grown Salmon, Institute for Retail Studies
    Market Report No. 2, University of Stirling
    1985.
5   P J W N Bird, Econometric Estimation of World
    Salmon Demand, University of Stirling
    Discussion Paper in Economics No. 111, 1984.
6   Development Planning and Research Associates
    Inc., in association with Frank Orth and
    Associates Inc., Alaska Salmon Projected 1982
    Market Conditions. Study carried out for the
    Department of Commerce and Economic
    Development, State of Alaska 1982.

REFERENCES

Earley, J V, Biing-Hwan Lin, Johnson R S, Whittaker, G <u>Alaska Salmon Prices and International Markets</u>, with DPRA and Frank Orth for Office of Commercial Fisheries Development, Department of Commerce and Economic Development, State of Alaska 1983

Biing-Hwan Lin, <u>Econometric Demand Model for Canadian Canned Salmon and an Econometric Model for North American Canned Salmon</u>, University of Alaska, Fairbanks 1983

Johnson R and Wang, D H <u>Markets for Canadian Salmon: An Economic Analysis of Market Demand</u>, Ottawa, Canadian Department of Fisheries and the Environment, 1977

Johnson, R and Wood, W R <u>A Demand Analysis for Canned Red (Sockeye) Salmon at Wholesale Level</u>, Special Report No. 411, Oregon State University 1974

Chapter Nine

SUMMARY AND OVERVIEW

9.1 OUTLINE

The preceding chapters have ranged widely over the various sectors and relationships in the salmon business. This final chapter summarises findings and discusses key parameters which will determine the pace and nature of the developments of the industry over the coming decade.

9.2 THE SALMONS

Salmon are found in both Pacific and Atlantic waters. In terms of volume and value the Pacific salmons are the most important, accounting for over 660,000 tonnes or 96% of total world landings in 1983. There are five commercially significant types of Pacific salmon: chinook (ONCORHYNCHUS TSHAWTSCHA), chum (ONCORHYNCHUS KETA), coho (ONCORHYNCHUS KISUTCH), pinks (ONCORHYNCHUS GORBUSCHA), sockeye (ONCORHYNCHUS NERKA) and lastly, cherry (ONCORHYNCHUS MASU) which is found in small quantities in Japanese waters. Pinks and chums are the most abundant of the Pacific salmons but chinook, sockeye and coho normally sell at higher prices.

All the nations with Pacific coastlines from the temperate zone northwards are producers of salmon. The largest producer is the United States which is responsible for about 45% of the total, and most of this comes from its Alaskan fishery. Japan produces about one quarter of the total with the remainder coming from Canada (British Columbia) and the USSR. Small but increasing quantities are coming from the Southern Hemisphere as a result of successful stock transplantings. With the exception of a small but growing volume of farmed salmon, Pacific salmon sold commercially are wild salmon captured by various fishing methods from the ocean waters in which they have been ranging freely. However, in order to increase the available supply many of these salmon come from hatchery production rather than from wild stocks.

The salmon found in Atlantic waters are Atlantic salmon, SALMO SALAR. Atlantic salmon spawn in the rivers and streams of the countries of Northern Europe, Canada and the United States and migrate to sea feeding grounds off areas such as Greenland, Iceland and the Faroes before returning to their native rivers to spawn. They are found in much less abundance in the wild than Pacific salmon. Indeed, due apparently to over-exploitation and deteriorating environmental conditions, wild stocks are now considerably lower than historic levels.

A decade ago it looked as if in commercial terms, Atlantic salmon were being increasingly relegated to a specialist small volume niche in international markets. This scenario has now changed because of the rapid development of Atlantic salmon farming in the 1970s and 1980s, mainly in Norway but with important developments in other Northern European countries, especially Scotland, and with some developments in North America. Farmed salmon are held throughout their entire life cycle in captivity, usually in cages in seawater during the on-growing stage. The significance of salmon farming lies not only in its rôle in increasing supplies of Atlantic salmon. Equally important, it represents a revolution in production technologies since salmon farmers have the potential capability to produce salmon to the size, weight and physical specifications required by the market. The timing of harvesting can also be adjusted to meet market needs and is not governed by the patterns of wild harvesting seasons. The largest quantities currently farmed are of Atlantic salmons but this development has been so successful that there are now increasing levels of production of farmed coho and chinook in the Pacific salmon producing countries.

9.3 FRESH AND FROZEN SALMON

Substantial quantities of salmon are marketed in fresh form. The amounts depend on market preferences, the location of harvesting points in relation to markets, the volumes involved and the relationship between the timing of harvesting and patterns of demand. Traditionally, when harvesting points are close to markets and adequate means of transport are available there is a preference for marketing salmon fresh because freezing or canning costs are avoided and customers are generally prepared to pay premium prices for fresh fish. The scope for the marketing of fresh salmon more widely has however increased considerably in recent years. One of the most interesting developments has been in methods of transporting fresh salmon over long distances. For instance, air freight methods have been developed for the freighting of farmed salmon from Norway and Scotland to other European markets, to North America and even to Japan. Increasing quantities of fresh salmon are also being transported by air from Alaska to other parts of the United States. This has greatly extended the markets for fresh salmon throughout the world.

Salmon are frozen when there are imbalances between the timing of harvesting and consumer needs. Much of the North American and Japanese catches are frozen because of the very seasonal nature of landings and because the large volumes caught in short periods of time would be difficult to handle in any other way. Given the distance of some harvesting points from major markets, particularly in Alaska, it is the only way in which non-canned salmon can be transported to markets at moderate costs, of some importance since not all markets are prepared to pay the higher prices associated with the high costs of air-freighting salmon. Most farmed salmon are sold fresh because on the whole production can be planned in line with patterns of demand and is not dependent on the natural migratory cycle. Exceptions are the freezing of seasonal surpluses, the freezing of consumer convenience products and freezing for some smokers who prefer to buy salmon in this form.

The amounts of processing involved, even for frozen salmon, are relatively small and fairly simple. The barriers of entry for new firms into handling and processing are low because the costs of equipment are low and equipment can be hired and because production economies of scale are not very large. As a result a wide variety of different sizes of operation is found. Sometimes the fishermen and farmers do their own marketing either individually or through co-operative marketing arrangements. Alternatively, the fish may be sold to specialist processors who prepare and freeze and make contact with customers. Nevertheless, although there are many small companies, the bulk of the volume is handled by a relatively limited number of large companies and in North America these larger companies are also heavily involved in the canning of salmon. In the absence of significant production economies of scale, their size partly reflects their ability to buy well and to market their products effectively. In addition some of their customers, especially salmon smokers and large wholesalers require such large volumes of product that only companies commanding a large volume of supplies can meet these needs on a regular basis.

Markets for fresh and frozen salmon have been growing. One reason for this growth has been rising incomes in consuming countries but sales have also been helped by better marketing of products through improved quality control and more effective presentation and promotion. The largest consumer of fresh and frozen salmon is Japan which has a large domestic supply, mainly of chum salmon and imports very substantial quantities from North America. The other major markets are in the United States, Canada, France, West Germany, the United Kingdom and a number of other European countries. The United States and Canada are large markets for their own salmon and for farmed salmon from Europe. European markets consume Atlantic farmed salmon and import substantial quantities of Pacific salmon.

9.4 CANNED SALMON

Canned salmon is synonymous with Pacific salmon: virtually no Atlantic salmon are canned at present. Japan, the USSR, the United States and Canada all have salmon canning industries but production from the United States dominates. In the United States around 300,000 tonnes of salmon are canned each year, a figure which represents over 40% of United States landings. The most popular species for canning are sockeye (red), pinks (pink) and chum salmon. Salmon canning is a sophisticated business using expensive equipment and requiring the highest levels of quality control. Salmon are delivered whole to processing plants located near to the major landing points. Here the roe are extracted and the fish are headed, gutted, cut into pieces and put into cans which are then sealed and cooked. Because of the high capital costs of equipment and the existence of some economies of scale, the salmon canning industries in North America exhibit moderately high levels of concentration. A further explanation for this concentration is that since they are buying very large quantities, they may gain some advantages in the prices at which they buy. Larger size also gives them the same advantages in selling that were mentioned above in the fresh/frozen salmon case.

Some of the salmon canners own their own fishing fleets to guarantee supplies to keep production lines going. In recent years, however, particularly in British Columbia, this has caused problems. The low profitability of the salmon fleet and its high debt burden have made such ownership a doubtful asset. In any case, however, the fortunes of canners and the fishing fleet are inextricably linked. Canners provide loan finance for fishing fleets, even for boats which they do not own and in return many fishermen pledge their supplies to the canners.

Canada and the United States have large domestic markets for canned salmon, as well as being the largest exporters. Other exporters are Japan and the Soviet Union. Exports go to a variety of markets, the largest of which is in the United Kingdom. There are other substantial markets in France, the Netherlands, Belgium and Australia. Canned salmon markets have been fairly static in recent years, as have per capita consumption levels, partly because rising incomes do not stimulate demand for canned salmon to the extent that they do for fresh and frozen salmon. Markets for canned salmon are also very price sensitive which has caused particular problems for North American exports because of rises in their international currency values. Nevertheless improved marketing in recent years has increased returns and there is currently optimism about the future.

243

9.5 SMOKED SALMON

Probably 60 to 70% of Atlantic salmon, and around 30% of Pacific chinook, coho and chum reach the final consumer as hot or cold smoked salmon. The smoking process starts with fillets of salmon which are chilled in a brine solution before being smoked either at low temperatures for long periods to produce cold smoked salmon or at higher temperatures to produce a more strongly flavoured "kipper" type product. A variety of different types of flavour can be produced depending on the type of smoking materials used. The result is a high value product which, unless frozen or held at chill temperatures has a relatively short shelf life.

There are major salmon smoking industries in the United States, in the United Kingdom, France, Denmark, West Germany, Norway and the Netherlands. On both sides of the Atlantic, both Pacific and Atlantic salmon are smoked, often within the same establishment. A considerable proportion of the frozen salmon exported to Europe from North America is in fact destined for salmon smokers.

There are markets for smoked salmon in North America and in Europe, but only as yet a small market in Japan. Small quantities are however exported to a very large number of countries worldwide, which is consistent with smoked salmon's status as a luxury product going to narrow markets. Most of the countries with major smoking industries, with the exception of West Germany, are also substantial exporters of smoked salmon. Some of the frozen salmon exported from the West Coast of the United States to Europe for smoking even finds its way back to the United States subsequently.

Salmon smokers vary in size and in the extent of their other interests outside smoking. This is because production economies of scale are not large and because small smoking units can be bought at moderate cost. In the United Kingdom, as an example, smokers range in size from annual production levels of 1,000 tonnes down to 5 or 10 tonnes. Much of the export business however is in the hands of the larger companies.

Markets for smoked salmon, particularly cold smoked salmon, have been growing rapidly. There are no precise figures available but the growth in demand is likely to have been stimulated partly by rising incomes and partly due to changes on the supply size associated with changing technologies which have made smoked salmon available to a much broader range of consumers than before. These changes have introduced smaller consumer packs of smoked salmon and vacuum and controlled atmosphere packs which have extended the shelf life of chilled products. Because of these changes a wider range of retailers have become interested in stocking smoked salmon and increasing numbers of consumers have found the smaller packs affordable.

9.6 DISTRIBUTION AND MARKETING

The handling and marketing of salmon products after they leave the harvesters and processors is just as important to the salmon industry as the earlier stages because of the effects on the quality of the product which final consumers receive and the prices which they pay.

Often marketing channels are long, involving a number of intermediaries between the harvesters, processors and the final consumers. This is most likely to be the case when products are traded internationally and when either or both harvesters and retail and catering customers are small. On the whole, however, there has been a tendency for marketing channels to shorten in recent years due to the increasing size of many retailers and caterers. The large volumes which the latter buy make direct dealing with processors and harvesters worthwhile.

On the whole, there is little formal vertical integration in marketing channels apart from the changes mentioned above. Harvesters and processors are not integrated forwards into general fish or food product wholesaling. This is left to specialist wholesalers who distribute a wide range of fresh fish products, frozen food or canned food products. Nor has there been a great deal of international integration across national barriers, although there is now an increasing number of exceptions to this. The first major exception is the integration of many Japanese fishing and trading companies into the processing of fish in North America to complement their domestic operations which already encompass fishing, processing and trading. The second and more recent change is the growing number of Norwegian farming and trading companies who are

extending their interests in farming and trading
into other European countries and into North
America.

9.7 SUPPLY ISSUES AND THE FUTURE DEVELOPMENT OF THE INDUSTRY

The profitability and development of salmon
markets of course rests on an adequate supply
base. The adequacy of the supply base however,
as earlier chapters have shown, depends
increasingly on intervention by man into the
management of the salmon resource.

Salmon enhancement and 'ranching' has in the
majority of cases to date involved the
replenishment of existing stocks, particularly in
the Pacific coast, which would otherwise have been
depleted through extensive exploitation and
man-made environmental and habitat changes.

The significance and the overall economic
benefits of enhancement, in which hatchery and
release costs would be measured against
incremental stock return, have been difficult to
assess, particularly in that these stocks often
add to existing wild stocks in the same freshwater
systems, and at sea mingle and are caught all
along the return migration route. The overall
evidence however is that if properly conceived
enhancement can be a cost-effective tool in salmon
production, but that results can vary
significantly, depending on any or all of the
factors influencing the relationship between
hatchery cost and return value.

In the case of ranching, which in ideal
circumstances does not conflict with existing
stocks and their associated fishing rights and
legal constraints, the economics of operation can
be far more clearly defined. In these
circumstances, a positive biological return is
likely to increase overall salmon production, and
tend to supplement overall stock levels.
However, at present, experiences in commercial
salmon ranching have been limited, and the results
very variable. At one extreme, the ranching of
pink salmon by the Alaskan Aquaculture
associations has been extremely successful; at the
other, the Pacific salmon ranching operations in
Oregon and the Atlantic salmon trials in Scotland
have shown limited results.

At present, because of the limited range of

ranching systems, and the very dominant effects of
site choice, it is difficult to define significant
economies of scale, though hatchery production
cost, the most directly controllable cost element,
is reduced at larger scale. More significant is
the biological difference between the stocks which
requires that Atlantic salmon spend 18 or more
months in freshwater, while Pacifics may be
released at 5-6 months. The resulting difference
in production costs means that Pacifics require
far fewer adult fish to return before yielding an
economic return, while it is difficult to ranch
Atlantics viably.

In contrast with enhancement, the production
of farmed salmon, though potentially more
expensive, uses external husbandry and feed
resources to remove salmon production from the
constraints of the natural freshwater and oceanic
productivity. In doing so, many of the
traditional legal, commercial, and market patterns
are subject to change, and are shown to do so.
Subject only to the technical factors affecting
production cost, the salmon aquaculture industry
has expanded, and to date continues to expand, at
a substantial rate, involving many geographical
areas and many trade and distribution systems
which were not previously associated with salmon
production.

One of the most important characteristics of
salmon farming has been the ability to supply fish
in a relatively controlled manner throughout most
of the year, thereby permitting the development of
entirely new markets and products. It is this
characteristic which has assisted the growth of
the industry.

The commitment to production levels must
however be made some time before actual output.
This is reinforced by the relatively high levels
of indirect cost, which make it difficult for
operators to reduce capacity significantly in
response to changing demand. There is thus the
potential for shorter-term instabilities in the
demand and supply relationship, caused either by
unexpected changes in oceanic production, or
over-optimistic investment in farming capacity and
stock. In these circumstances, though some
protection may be available through market
segmentation, operators of higher-cost methods of
production may be at significant risk, and some
capacity adjustments may be required.

9.8 EXTERNAL INFLUENCES AND THE SHAPE OF THE INDUSTRY

9.8.1 Rôle of Governments

How the changing supply situation interacts with demand opportunities has been discussed in Chapter 8, where it was noted that careful marketing should ensure that opportunities exist to develop markets to absorb the increasing levels of output. There is at the same time a number of external factors which will affect the shape of the industry in response to these varying patterns and the relative gains and losses of different sectors. The first of these is the extent to which governments, individually or collectively become involved in the management of the industry. In the circumstances where potential oversupply conflicts may occur, governments may be inclined to take an active rôle in protecting national investments and stabilising markets. This would represent a significant departure from the main aim to date: that of preserving or building up salmon stocks. There would be several means available for this: the use of licences, and the control of either fishing effort or aquaculture production capacity, the use of import controls, price regulation, or the use of support price structures.

In practice, these actions would have to be considered very carefully; Chapter 3 outlines some of the practical constraints to the use of regulation of fishing effort, while attempts to license or control aquaculture development may be inequitable and politically problematic unless carefully conceived. Given the difficulty of rapid adjustment to changing market conditions, the use of price controls or support mechanisms could carry with them the danger of sustaining for the longer term the sectors of production which are otherwise nonviable. At the very least, protectionist measures, whether local (e.g. for remote areas), national, or regional (e.g. EEC) would distort trading and production patterns, and could act against the longer-term efficiency of the industry.

9.8.2 The Internationalisation of the Industry and its Impact

Any actions taken to control, regulate, or support domestic production would increasingly have to take account of the growing internationalisation

of the industry: there are increasing numbers of
supply points, and increasing areas of market
demand, using a wider range of currencies. The
competitive effects, and the effects of currency
movements can have significant impact on national
policy and the competitiveness of supply sectors
within each country. Currency movements, in
particular, can have a pronounced destabilising
effect on supply relationships.

A further factor to be considered is the
increasing internationalisation of companies
within the salmon industry. Thus several
Norwegian companies are active in Europe and in
North America, Japanese groups have invested in
processing capacity in North America and in South
America, and many others are actively planning
multinational activities. While this gives the
individual companies greater strengths through
diversification, their effects on international
transfer prices can create distortions in simple
comparisons of production cost and competitiveness, and their position may make local regulation
difficult to implement. Within any national
context there would also be considerable potential
for conflict between national and multinational
operators.

9.8.3 Technical Developments
Improvements in production efficiency through
technical development have the potential of
reducing production costs and hence changing
supply and demand relationships. As shown in
Chapters 4 and 5 it appears unlikely that
significant technical changes will occur, though a
gradual improvement of efficiency through improved
feeds and feeding systems, genetic control, and
simplified production systems, may be expected to
stabilise, or even reduce costs slightly under
favourable conditions. There are thus certain
production price thresholds below which it is
unlikely that producers can operate, with a range
of production prices above these limits,
corresponding to site and system constraints.

The rôle of technical change in the processing
and distribution sector may be more significant.
Improvements in the methods of presenting fresh or
freshly prepared salmon products may be expected
to be crucial in the expansion of markets, and may
also have significant effects on the requirements
for harvest, distribution and storage of primary
salmon products.

9.8.4   Recreational Fisheries

Although this subject has not been considered in detail in the course of the book, there are several issues of importance in the longer term, where demand for recreational fishery, and its increasing economic and political significance, can affect potential for supply, and to some extent, local markets.   Perhaps the most important issue is that of competition for fishing rights between commercial salmon fishers and recreational fishers.   In the case of river fisheries, the issue is normally that of limiting estuarine and coastal commercial fisheries, to allow sufficient numbers of fish to return to river areas for sport fishing and for spawning.

In the coastal recreational fisheries such as those on the west coast of North America, commercial fishing inshore may be completely supplanted by sports fishing.

In most cases studied, value attributed to fish caught through sport can be shown to be several times that of the commercial value of the fish, and there is thus a powerful economic argument for allocating stock resources preferentially to sport fishing.   There are clearly limitations: vastly increased numbers of fish being allowed to run up river may not substantially increase economic returns to a form of fishing whose value is strongly related to exclusivity.   Ideally, management would seek to maximise returns on the higher value fishery, developing it to the point where marginal returns reached those of the lower value commercial fishery.   What is not clear at present, however, is how this relationship will be quantified if the sports fishery sector is substantially expanded.

It is however the case in many areas that the recreational sector is expanding, and in many instances has a considerable political importance.

In the longer term, this effect may be expected to limit the potential of the traditional fisheries, and in some cases, to constrain the potential of ranching.   In the case of enhancement, the strength of the sport fishing lobby may encourage development, and suitable locations may result in positive benefit to both commercial and recreational sectors.   Salmon farming, by contrast, is unlikely to be directly affected, though any activities considered to interfere with sports fisheries, such as freshwater control by hatcheries, selection of specific stocks and/or their accidental release, or occupation of inshore fishery areas, could cause conflict, and could result in constraints to farmed production.

The effects on markets of greater sport fishing catches are difficult to assess. Given that most such catches are seasonal, the year-round potential of salmon farming is unlikely to be affected, and although the households supplied with sport-caught fish may reduce purchases of salmon through conventional markets, this deficit may easily be made up by a wider appreciation and demand for salmon in out-of-season times.

9.8.5   Environmental Issues

There is a number of important environmental issues concerned in the longer-term development of the salmon industry. For the wild, enhanced, and ranched stocks, the conservation of the freshwater habitat is one of the most apparent environmental needs. In the marine environment, the effects of ocean current changes, feed resources, and coastal pollutants can have considerable effects. There is increasing evidence of the sensitivity of salmon to pollution and environmental change, and whether or not these are controllable by human intervention, their effects could be significant in changing supply and quality.

For aquaculture of salmon, there are increasing concerns about environmental impact, both visually and in terms of nutrient-rich waste products, which may in fact risk "self-polluting" the production sites. At present, there are several technical programmes underway attempting to tackle this issue; clearly if environmental impacts become too severe, there will be a substantial constraint to production. If control measures prove particularly costly, and apply differently in different areas, relative production costs can be expected to change.

9.9 THE FUTURE STRUCTURE OF THE SALMON INDUSTRY

As the relevant sectors of the text have shown, there are many factors determining the present structure of the industry, and many factors which might be expected to change and shape the industry of the future. It is clearly impossible to predict exactly the future characteristics of the industry, but it is useful to comment on the significance of the major trends, and how they might act.

In the fishery sector, it would appear that the rôle of licensing and control of fishing effort will continue to influence individual catching unit size. In certain cases, the availability of other fisheries may be more important in determining vessel type and size. At present, most legislation is concerned with supporting a reasonable number of smaller units, and such is the nature of the fishery itself, it is probable that moderately sized fleets of smaller vessels may be the most effective and sensitive means for operating the fishery. It is thus unlikely that significantly larger catching units will be employed. The organisation of the vessels within the industry may change, however, either through increased co-operation and shared use of facilities, or increased concentration of ownership. However, as Chapter 3 points out, the low or variable profitability of many vessels makes them unlikely targets for purchase unless specific supplies need to be secured. As shown, there do not appear to be any particular incentives for this.

A further issue affecting the fishery sector is the rôle of governments in restructuring the industry as a whole. In many of the areas involved with salmon fisheries, grants and subsidies are available to decommission vessels and/or to surrender licences. There may thus be a longer-term trend towards reducing the number of participants in the industry.

In the case of enhancement or ranching operations, the rôle of co-operatives, formed either from fishery operators, sports fishing interests, or other community groups, is likely to increase in importance, and may in some areas take over the functions of public enhancement programmes. As shown, in Chapters 4 and 5, smaller scale hatcheries in suitable locations are potentially viable, particularly if supported by voluntary community labour. However, the operation of true "ranching" systems is likely to be confined to larger operators, sometimes combining with hatchery production for salmon farming, who can afford the risk, and long lead· time of development. Any such operations are therefore likely to be at a larger scale, sufficient to support fully commercial overheads.

In the farming sector, the effects of government control in regulating area or output are likely to continue to dominate unit size, though wider distribution of ownership may occur, with investors holding shares in several sites within one or more countries. In countries such as Scotland, Iceland, and Ireland, where no specific limitations are imposed, maximum unit size will tend to be controlled by site capacity, and the desire to reduce disease transfer risks. In farming, as Chapter 5 discussed, the effects of individual site factors will often outweigh specific economies of scale, so it is probable that a considerable diversity of unit size will remain. This will be assisted by the development of co-operative activities in e.g. stock supply, feeds, marketing. For larger companies, the trend will be to increase the number of sites, and to centralise management services and infrastructure.

In the processing and distribution sectors, as Chapter 6 shows, there are no dominant forces towards larger unit sizes at production level. Smaller production units, often geographically dispersed, appear to be able to operate at similar unit costs. However, the changing rôle of larger multiple retailers and concentrations in the catering sector may influence the size structure of businesses towards larger suppliers, leaving smaller suppliers in the more specialist market sectors. Alternatively, if smaller suppliers wish to preserve their independence but take advantage of the larger market outlets, solutions may be found through the formation of marketing companies or through co-operative marketing arrangements.

The change in supply and distribution wrought by the salmon farming industry has been particularly influential in opening up new commercial opportunities within the industry.

Traders in wild fish tend to preserve their flexibility to move between customers and supplies due to the rapidly changing and unpredictable nature of market conditions. By contrast, there are some indications that farmed salmon may be causing changes in traditional relationships. Because of the greater predictability of supplies of farmed salmon and the ability to produce a product meeting specifications, prices are tending to show more stability. The advantages of flexibility are therefore less and the advantages

of building up close relationships between the
different stages are becoming more important.
Some use is being made of contractual fixed price
agreements between suppliers, processors and
retailers as one indication of this. Since this
is in line with modern business arrangements in
many Western countries this process is likely to
continue and possibly also extend into the
traditional wild sector. In some cases this may
lead to formal vertical integration from supply
through to distribution, processing and marketing,
although this is not expected to be a strong
trend. Although there are some examples of this
at all production levels, it is mainly the
smallest and largest scale operators who are
involved. At the small scale, there may often be
incentives to distribute, process and sell
locally, adding value to small levels of output,
while at the large scale, the size, existing
structure, and development objectives of the
companies concerned provide the motivation to
develop vertically. Against these trends,
however, must be counted the emergence of
specialist operators, whether supplying eggs or
smolts, distributing feeds or fish, processing
supplies, or marketing to particular sectors or
geographical areas. Processors for instance who
can assemble a wide product range of which salmon
products are only one item compete strongly with
more specialist salmon companies. Wholesalers
perform a similar rôle. The existence of such
strong non-integrated operators limits the
opportunities for vertical integration. The
possibility of achieving benefits of co-ordination
through strong informal relationships may also
reduce the need for formal vertical integration
and its consequent lack of flexibility.

9.10 CONCLUSIONS

The salmon industry is a complex and diverse
activity, covering a wide geographical area, and
becoming increasingly international in its
characteristics. One of mankind's last wild food
resources, salmon are also increasingly supplied
through controlled husbandry. The varied sources
of supply, with their varying efficiencies, costs
of production, and supply characteristics,
interact with an equally varied system of
processing, distribution, and marketing. The

changes in the salmon industry, and the many factors contributing to these changes, have considerable implications for the future structure of the industry, on the range of supply available, and on the operation of the market. The impact of farmed salmon production has been and will be significant; its effects, the effects of the increasing environmental and resource pressures on the traditional sources of supply, and the changing patterns of product availability and market demand are likely to continue the process of change in the industry. This text has attempted to describe the industry, to explain and account for its characteristics, and to identify the ways in which it might develop in the future. These are exciting times for the industry; quite how it evolves will depend much on a full understanding of the way it works: it is hoped that this text has gone some way to contributing to this understanding.

AGENTS: dealers who sell on a commission basis for processors or harvesters

ALEVIN: Stage of development of young salmon after yolk-sac absorption

ANADROMOUS SPECIES: salmon returning to source of origin to spawn

BRAND NAME: when the name of the seller is prominent on the packaging and clearly identifiable to the buyer

BRIGHTS: unlabelled cans of salmon

BROKERS: buyers and sellers who facilitate deals but who do not usually handle the goods themselves

CANNED SALMON: salmon cooked and sold in airtight containers

COMMERCIAL FISHERIES: catching of fish for re-sale for consumption

CONTROLLED ATMOSPHERE PACKAGING (CAP): use of gas mixes around the product to prolong shelf life

COST, INSURANCE AND FREIGHT (CIF): cost of delivery to customer

DEMAND SCHEDULE: response of demand to different prices

DRESSED SALMON: salmon which have the head and gills removed and are eviscerated

DRIFT GILLNETTING: fishing using suspended nets which trap the on-coming fish in the net

EEC: European Economic Community

ELASTICITY OF DEMAND: the measure of the response of demand to a change in another variable, particularly changes in price and changes in incomes

FAO: Food and Agriculture Organisation of the United Nations

FARMING: the rearing of fish in captivity during some or all of their life cycle

FREE ON BOARD (FOB): prices excluding delivery costs

FRY: General term for young fish, salmonids in the early freshwater stage

ECONOMETRICS: the application of mathematical and statistical techniques to economic problems

ENHANCEMENT: the release of artificially reared stocks for eventual capture through the operations of public fisheries

EX-VESSEL PRICES: price received at the dock or point of transfer from catching or holding vessel

GRILSE: Atlantic salmon returning to freshwater after one sea-winter

GUTTING: evisceration of fish

HARVESTING: this term has been used to cover both the capture of salmon through fishing and through farming

ICES: International Council for the Exploration of the Sea

INTERCEPTION FISHERY: where fish are caught on their return migratory route but some distance from their native freshwater areas

LANDINGS: fish brought ashore from fishing operations

LIMITED ENTRY FISHERY: where entry into the fishery is controlled, usually by licensing systems

MARKETING CHANNEL: the route taken by products from their source to the final consumer

MILT: the spermatozoa and seminal fluid produced by a fish

MIXED STOCK FISHERY: one where fish from different river systems are mingled

NMFS: National Marine Fishery Service of the United States

PACKS: the containers in which cans of salmon are packed and stored

PUBLIC SALMON FISHERIES: where fishermen have rights to harvest but not rights of ownership

RANCHING: the release of artificial stock for eventual recapture close to the release point, usually for private commercial gain

REAL INCOMES: incomes adjusted for the effect of price inflation

ROE: the mature eggs of salmon

SEINE FISHING: fishing with suspended circular nets

SMOKED SALMON: salmon cured by smoking

SMOLTS: the young of some types of salmon at the stage where they migrate to the sea

SPAWNING: the production of eggs

SPORT FISHING: fishing as a recreational activity rather than for commercial gain

SUBSTRATE: non-living material e.g. sand, gravel on which salmon live

SUPPLY SCHEDULE: a schedule of quantities supplied at various prices

TAC: total allowable catch

TERMINAL FISHERY: where salmon are caught at the end of their migratory route

TRAPS: methods of catching salmon at their return to release points and spawning areas

TROLL FISHING: fishing using lines and baited hooks trailed from small fishing boats

VERTICAL INTEGRATION: the joint ownership of successive stages in processing, fishing and marketing

WHOLESALER: buys and resells products to other members of the distribution channel but not to the final consumer

WHOLESALE SELLING PRICES: the prices at which wholesalers sell

INDEX

fishermen 169, 172-3, backward integration
171-2, process diversification 167-8
Cash-flow characteristics, ranching 104, 106,
effect of early returns on 89; aquaculture
139-40
Catch fees 16
Cherry 6; world production 24; ranching and
enhancement 82,90
Chile 2; aquaculture 18; ranching and enhancement
82, 90, developments 109; 114
Chinook, 6; life cycle 6-8; product form 8, 20,
32, 44; fishing methods 11; supplies 24, 26
and sport fishery 29; % canned 32, 33;
conservation 73, enhancement and ranching 82,
86, 90-1, and return rates 87, hatchery
facilities 92 and minimum release numbers 94;
aquaculture 113-4
Chum, life-cycle 8-9; supplies 23, 24, 25, 26,
29; product form 9, 20, 32-3, 44, 48, 54;
stage caught and quality 68; enhancement and
ranching 82, 83, 90-1 and return rates 87,;
hatching facilities 92; minimum release
numbers 94; future markets 233
Coho 9, 11; product form 9, 20, 32, 33, 44, 45;
supplies 24, 26, 29; enhancement 82; ranching
90, return rates 87, 95, and minimum release
numbers 94; aquaculture 113, 117, and
production costs 119; future farmed markets
232-3
Columbia River Project 82, 84
Conservation, need for 55-6, 67; and incentives
55, 67-9; treaties 71, 73.   SEE ALSO Controls
Contamination and multiple sites 126, 145; and
recycled water systems 145
Controls:- see Area licensing, Area/time closures,
"Buy-back" programmes, Gear bans, Davis Plan,
Licensing, Limited Entry, Quotas
Cooperatives and conservation incentives 68; in
marketing 174-5; future rôle in enhancement 252
Curing, types of 175.   SEE ALSO Smoking
Currency fluctuations 155; and effect on prices
226-7

Davis Plan (Canada) 61
Demand, effect of changing prices 213-4, 222-3;
and substitutions 218, 222-3, 227; and income
levels 224; and exchange rates 226-7, and
trade barriers 227-8; and consumer preferences
228; elasticity of 236, 257, as limit to
growth 231, and as rôle in investment 214
Demand analysis 215

171, and effect of buyers' market 172;
forward, and as marketing control 173;
international 157, 197, 245 and effect on
regulations 249; vertical 172, 181, 182, 259,
and industry future 182, 254
Interception fisheries, effect on landings 25, 29;
rights and conservation 68, 69, 71, 73; 258
International integration 157, 197, 245 and
effect on regulations 249
Investment, aquaculture 143-4; and return 232;
and elasticity of demand 214; and future of
industry 254
Ireland as supplier 30, 32; as market 38;
controls 59; enhancement 82; ranching, return
rates 87; aquaculture 114 and controls 119-20
Italy as market 39, 41

Japan 6, 8, 13; aquaculture 18; supplier 25, 29,
38; market, roe 20, 48 and farmed salmon 43
and smoked salmon 48; canned trade 33, 35 38,
44; fresh/frozen trade 33, 41-2; conservation
68, 73; enhancement and ranching 79, 80, 82,
83; freezing 154; integration 157, 171, 197;
distribution channels 189, 197-8; price
elasticity and substitutions 223; 239

Kings - SEE Chinook

Labelling - SEE Marketing
Land-based fishing 13-15
Landings and process choice 21, 35; and
production by species 24, 25, 29-32 (Atlantic)
and 25-9 (Pacific)
Licensing, area 59, 67; limited entry 59, 61-2,
63, 64, 65, 67 and backward integration 171
Longlining, SEE trolling
Luxembourg, as market 38, 41, 43

Manchuria 6
Management, objectives 54-9, and conflicts 57-8,
61-2; and control difficulties 54-75;
government rôle 248-9; and future of industry
252.   SEE ALSO Aquaculture, Conservation,
Controls, Interception Fisheries, Regulations,
Treaties
Marketing fresh/frozen salmon 154-5, and
promotion 155-6, and costs 161;   canned salmon
165-6; smoked salmon 175, 178-181; and forward
integration as control 173; effect on demand,
155-6, 166, 178, 228-9.   SEE ALSO
Distribution channels, Generic marketing,

265

Integration